The Philosophy of Zoology Before Darwin

The Philosophy of Zoology Before Darwin

A translated and annotated version of the original French text by Edmond Perrier

Originally published by Félix Alcan, Paris in 1884

Translated by Alex McBirney with annotations by
Stanton Cook and Gregory Retallack
University of Oregon

Springer

Alexander McBirney
Department of Geological Sciences
University of Oregon
2780 Capital Drive
Eugene OR 97403
USA
mcbirney@smtp.uoregon.edu

Stanton Cook
Department of Geological Sciences
University of Oregon
2780 Capital Drive
Eugene OR 97403
USA

ISBN 978-90-481-3008-5 e-ISBN 978-90-481-3009-2
Springer Dordrecht Heidelberg London New York

Library of Congress Control Number: 2009930626

Printed on acid-free paper

Springer is part of Springer Science+Business Media (www.springer.com)

Table of Contents

Foreword

Jean Octave Edmond Perrier was a French zoologist who lived through the tumult of British Darwinism and Lyellism, and reminds us in this revealing account that French scientists had much to contribute to such perennial topics as evolution, catastrophism and creationism. While very much a product of the Third Republic, Perrier's account also aimed to outline timeless issues and permanent advances in taxonomic and developmental biology since classical Greece and Rome. In this aim he succeeds with surprisingly modern perspectives for a book first published in 1884.

Perrier was born May 9, 1844 at Tulle, the son of the principal of a school which now bears his name, *Lycée Edmond Perrier*. In 1864 he was accepted to the *École Normale Supérieure*, where he was strongly influenced by Louis Pasteur and Henri de Lacaze-Duthiers. After working for three years at a high school in Agen, he obtained a post of naturalist-aid at the *Muséum National d'Histoire Naturelle* (1868), advancing in that institution to Chair of Natural History of Molluscs, Worms and Corals (1876–1903) and then Director of the museum (1900–1919) and Chair of Comparative Anatomy (1903–1921). Previous directors of the museum included many of the scientists he discusses in this book: George Cuvier (1822–1823, 1826–1827, 1830–1831), Isidore Geoffrey St Hilaire (1860–1861), and Alphonse Milne-Edwards (1891–1900). Perrier's own research on echinoderms and earthworms took him on several expeditions in 1880-1885, mostly to Atlantic and Mediterranean coasts, but also to the Caribbean. He was president of the *Société Zoologique de France* (1879), elected to the *Académie des Sciences* (1892) and president of the *Société Nationale d'Acclimatation* (1901–1921). He died in Paris, July 31, 1921, an established insider of French science.

Perrier was an early convert and evangelist for Darwinian evolution. He also wrote a biography of Jean-Baptiste de Lamarck, whom he viewed as a cofounder of the theory of evolution, but whose brilliant intuitions had been stifled by the powerful Cuvier. In 1908 a statue of Lamarck was erected at the entrance of the museum under Perrier's direction. In this book Perrier demonstrates a complexity and depth to Lamarck's work that belies the common British notion of Lamarkism as the failed null hypothesis of inheritance of acquired characteristics. Lamarck's own writings are equivocal on the issue of acquired versus inherited characters, as are Darwin's, written well before Mendel and modern genetics. This topic remains relevant today, as misconstrued "Lamarckism" is resurrected for theories of viral genome inheritance and mosaic evolution.

Catastrophism was another doctrine commonly portrayed as pitting Frenchman Georges Cuvier against the uniformitarian Scotsman Charles Lyell. The real story, as Perrier reminds us, is that Cuvier was a consummate empiricist, distrustful of theories such as the homology concept of Etienne and Isidore Geoffrey Saint-Hilaire or Lyell's uniformitarianism. It is ironic that Cuvier has become a straw man for a general theory of abrupt geological change that was peripheral to his core interests in functional morphology. Nevertheless, catastrophism is now being resurrected to explain rapid changes in faunas such as those at the end of the Cretaceous, Paleocene and Eocene, which Cuvier simply observed from his studies of the Paris Basin. Asteroid impacts, flood basalt eruptions, and methane outbursts from sea floor clathrates and igneous intrusions are now widely recognized as agents of abrupt geological change.

Scientific creationism is nothing new, because it had nineteenth century proponents from both sides of the channel. Englishman William Paley reveled in the intricacy of living creatures as evidence of a divine creator, offering the timeless analogy of the watchmaker and his marvelous time piece. In this book, Perrier reminds us of a meticulous elaboration of that same view by Louis Agassiz, who also regarded inferiority of non-European races as divinely sanctioned. Agassiz was no maverick or crank. Well known today for his pioneering studies of fossil fish and discovery of continental glaciation, Agassiz was first a professor in his native French-speaking canton in Switzerland and later director of the Museum of Comparative Zoology at Harvard.

A recent reincarnation of Paley's watch analogy is "intelligent design", but these arguments are now swept away by a tidal wave of molecular biological evidence for evolutionary relationships.

Much of the charm of Perrier's book comes not from its topicality, which is a wonderful intellectual exercise, but from surprises of natural history. He was a man entranced by the details of mule breeding, colonial invertebrates, South American marsupials, and deep ocean salps. Perrier rode an unprecedented wave of nineteenth century natural history exploration and romanticism, which engaged not only scientists but also the literate public for whom this book was written. Arguments concerning organismal development and evolutionary relationships have since become more sophisticated, testable and mathematically rigorous. It is intriguing to see how complex modern elaborations have arisen from simple opinions competing in a world of ideas, and how much hinged on chance observation from wide experience of exceptional cases. Perrier reminds us of an age when beetle collectors

(Darwin) and fossil hunters (Agassiz) made timeless theoretical contributions. My suspicion and hope is that we still live in an age when individuals and their ideas matter, despite the alluring gloss of expensive instruments, large grants, computer graphics, and collaborative teams.

Gregory J. Retallack
Department of Geological Sciences
University of Oregon
Eugene, Oregon

Translator's Preface

A good number of books have been written about the pre-Darwin history of the concept of evolution, but Edmond Perrier's work is exceptional in its extensive scope, balanced coverage, and objective commentaries. The author had a personal interest in his subject. As director of the Muséum d'Histoire Naturelle, he was familiar with the work of his illustrious predecessors, Buffon, Lamarck, Geoffroy Saint-Hilaire, and Cuvier, who developed many of the basic ideas that are central to his story. His survey is by no means confined to France, however. He explores the work of naturalists of many nationalities, as well as the classical Greek, Roman, and Arabian scholars. In doing this, he gives us an excellent historical account of the arguments for or against evolution and natural selection, which most French and German naturalists of his time had already accepted. His main concerns, however, seems to have been the more fundamental studies that drew on paleontology, comparative anatomy, and embryology to establish a linkage from one species to another and thereby demonstrated a continuity of evolving life over a long span of geologic time. In this sense, the book provides a welcomed balance to much of the Anglo-centric literature that deals almost exclusively with natural selection, a concept that Perrier shows can be traced back to the time of Aristotle.

Perrier's book was brought to my attention by Philippe Janvier, a paleontologist at the Museum of Natural History in Paris where Perrier was one of his illustrious predecessors. I immediately saw that it was an exceptionally comprehensive and thoughtful historical account of the developing concepts prior to the publication of *The Origin of Species*. To my surprise, however, the book has been almost totally ignored and is rarely cited, even in books dealing with the same general subject. After reading a few pages, I saw the reason for this neglect. Much of the writing is convoluted and, in places, almost incomprehensible to the modern reader. Without fully appreciating the magnitude of the task, I decided it was worth translating it into a more readable form that would make it accessible to a wider audience. Fortunately, I was able to enlist the assistance of two colleagues, Stanton Cook, a biologist, and Gregory Retallack, a paleontologist. With their help, I have tried to convey the meaning and spirit of the author's text, and this has forced me to take considerable liberty with the original wording. The long sentences have been recast in a simpler form, and more current terminology is used to express some of the abstract concepts that were still in a state of flux at the time the book was written. Many of Perrier's references in the footnotes are incomplete and difficult to identify, but I have chosen to reproduce them in their original form while providing a separate list of references that should be more helpful. A bibliographical Index has also been added to provide a convenient reference to the many individuals cited in the text, and Professor Cook has compiled an extensive glossary explaining the terminology and how it has changed with time. The earlier works we have drawn on to interpret these evolving concepts are listed with the references immediately following the text.

Acknowledgements

In addition to the invaluable assistance rendered by my colleagues, Greg Retallack and Stan Cook, several persons have helped me interpret difficult parts of the text. These include André Comandon, Richard Desroche, and, of course, Philippe Janvier. Even with all this assistance, however, I have had to use my own discretion in interpreting much of the text, and if I have failed to render it with clarity and accuracy, I have only myself to blame.

Alexander McBirney

Preface

The evolution of ideas has much in common with that of living things. Ideas are usually born as humble thoughts hidden among long-standing concepts, and they may remain inconspicuous as they slowly develop because they are not distinguishable from their progenitors. Little by little, some will take on a character of their own and continue to grow stronger until they can survive and evolve alone. They then go on to engender other new ideas that may follow a similar course.

All the members of a single family of ideas do not necessarily share the same destiny. Some perish without having played any role whatever; they exercise no influence and provoke no change. Others that initially differ little from those that are destined to fail may go on to make important contributions to human thought. Everyone then thinks that these ideas are self-evident and that they recognized their importance long before they became widely accepted – even to the degree of claiming to be their parent. This is why it is almost impossible to write a history of ideas that everyone would consider impartial. It is why everyone who thinks he has an idea that is a new contribution to the store of human knowledge immediately finds himself assailed by the outcries of countless self-styled precursors to whom he had failed to give credit for early insights that supposedly played a role in fostering the development of these ideas.

For this reason, I have refrained from attempting to make this little book a complete survey of all the varied concepts to which zoologists have been led in their studies of animals. I have benefited from comments of fellow members on the staff of the Jardin des Plantes who have read some of the following chapters, but I leave it to historians to chronicle the minor events I may have neglected and to biographers to record the lives of the great men whose ideas I discuss. I have also deliberately avoided discussing all the nebulous, ill-conceived ideas that have left little if anything to posterity. I have chosen instead to focus more closely on those strong, vigorous ideas that have made lasting contributions to what has become modern zoological philosophy. I have discussed these ideas in the context of the times when they first became innovative intellectual concepts. This, I think, is the only way I could write a short book that would be clear, concise, and useful.

A few misguided French scientists have criticized French science and belittled the important role it has played in the splendid flourishing of biology that now radiates everywhere, even into the thinking of our political leaders. Despite what some may say, France has not – thank God! – remained isolated from the important developments in zoological philosophy. Few other countries have furnished so many scientists who have been able to lay out general concepts with clarity and exactitude. I have had the agreeable task of recording the manner in which this progress has been achieved, and I hope that I have done so with total fairness and impartiality, not only with regard to scientists of other countries but also to those of my contemporaries with whom I have had occasion to discuss these philosophical principles.

In dealing with the Philosophy of Zoology before Darwin, I have had to trace the ways in which many of our present ideas have grown out of earlier ones. This has led me to review the broad trends of modern biology, the aims it pursues, and the methods to which it must adhere to attain them. Some of these methods have yet to become guiding principles of modern biology.

Even if the general acceptance of transformism is accomplishing a profound revolution in the direction that naturalists aim their research and in their way of interpreting the significance of their observations, this revolution is still far from complete. Too often, the older method, which physicists have referred to a bit disdainfully as the *method of naturalists*, continues to introduce discord over fundamental principles and the corollaries ones derived from them. In studying the works of famous naturalists, one is struck by the great differences between their methods and those of physical scientists. These differences reside much less in the use of subjective observations instead of objective experimentation than in the customary practice of physicists who go from the simple to the complex in their efforts to relate effects to their proper causes. For many years naturalists confined themselves to *comparing,* while physical scientists were striving to *explain.* This has changed, however, and naturalists are now trying to explain the phenomena they observe. They have finally given up their endless appeals to metaphysics in the *natural science* they cultivate and, by a strangetwist of fate, they have ceded its true name to other sciences from which it has derived the methods from which it should never deviate. Until very recently, the unfortunate term metaphysics has been used in reference to the general principles underlying the relationships between living creatures. When Aristotle introduced his *principle of final causes*, which Cuvier made the central focus of all his work, he caused naturalists to search for the ways in which an external will had brought harmony into the world. Leibnitz's *principle of continuity* was used by Cuvier and his disciples, Linnaeus and Bonnet, who seemed to see no relationship between cause and effect in the phenomena they were trying to explain. To them, the inherent continuity and gradational variations between related organisms is simply a reflection of the intelligent design of the universe. Similarly, Etienne Geoffroy Saint-Hilaire could offer no reason for the *uniform plan of composition* that he saw in the animal kingdom other than some sort of mysterious relationship between living creatures and their creator. Cuvier did not challenge this; he strayed even further in this line of reasoning when he proposed the four basic branches to which all animal life could be assigned. As soon as he tried to go beyond the observed facts, he found himself drawn to the doctrine of final causes or to the hypothesis that the essential forms of animals pre-exist in the germ cells carried in their bodies. When Cuvier's closest disciple, Richard Owen, laid out his *theory of archetypes* and Louis Agassiz developed his ideas about *species* and *classification*, there was nothing mysterious in what they proposed: natural history was for them nothing but a record of the will of God.

As soon as one accepts this hypothesis - or any other form of metaphysics - it becomes very difficult to entertain any other view of the living world. The belief in the *fixity of species* came at a time when very little was known about the animal kingdom, and the knowledge we have since acquired has demolished this belief by showing that the idea of fixed species can no longer be sustained. The hypothesis is no longer consistent with the essential features of living forms nor with the relationships of these forms to their environment, and it can not be explained in terms of the simplistic generalizations naturalists used in the past. Dugès' *law of organic conformity* and Steenstrup's *law of alternating generations* were metaphysical statements of this kind; they defined the principles nature follows in perfecting its works. Even H. Milne Edwards, who rejected the idea of immutability, built his brilliant work on the so-called *division of physiological work* which implied that nature has effective means to transform creatures gradually into increasingly complex forms.

In the first half of this century, naturalists vainly hoped to avoid the discredited metaphysical processes by invoking a vague being they called *Nature* to explain the features that they described so beautifully. Some of the articles they wrote for encyclopedias and dictionaries were devoted specifically to this supernatural being. Nature is the Universe, it is God, and if it is not that, it is nothing. This Nature that was said to prevail everywhere offered no real explanations in the sense that physical scientists understand the word.

Explanations of the variations in related forms of life reveal a simple element that all share in common, and they show that these variations can be attributed to identifiable causes. In zoology, any scheme that takes man or any of the vertebrates as a point of departure from which other organisms have descended fails to offer an adequate explanation for these modifications. To try to "explain" the lower groups of the animal kingdom by means of concepts derived from studies that included only vertebrates is to retreat from the principles of modern experimental science. All the difficulties we still experience in defining the *individual* or *species* are artificial in the sense that they are of our own making. They result from a failure to abandon narrow concepts based entirely on earlier studies of animals or the higher forms of plants.

Thanks to improved means of investigation, it has now become possible to reduce living things to the elements that they have in common, namely their cells, and these, in turn, are defined in terms of the fundamental substance they have in common: protoplasm. Thus, it has been possible to establish a continuous chain between organisms composed of a single cell and those with millions. Embryogenesis demonstrates that even the most complex forms result from multiplication of a single initial cell, the egg. Thanks to improved methods of investigation, true explanations appear to be forthcoming, just as they have for physicists and chemists.

We may be justified in hoping that it will some day be possible to present the history of living creatures in an instructive form appropriate for the experimental sciences. I have taken a first step in this direction in my book, *Les colonies animales et la formation des organisms*, published in 1884. But to achieve this we must be confident that the origins of natural organisms can be found in their present form and strive to find evidence of the causes that relate complex phenomena to simpler ones. This method will enable us to find links between more extensive assemblages, and we should not be deterred by critics who attempt to make a hypothesis seem faulty by pointing out something it leaves unexplained or explains in a way that does not include every possible relevant fact.

I hope that this account of earlier thoughts will contribute to progress by showing where the true path to understanding may be found.

Edmond Perrier

Chapter I

Introduction

The earliest ideas about the place of animals in nature. Myths and philosophies of ancient people.

Throughout time, humans have sought to understand the origins of the living creatures they see around them and to find an explanation, crude as it might be, for the links between those animals and ourselves. Since the dawn of human consciousness, we have cast a curious eye on those restless animals that shared our place on Earth. Unable to account for the existence of these mute creatures, we have always marveled at their extraordinary instincts, daunting power, charming colors, graceful movements, and elegant forms. At first, we saw them as agents of the invisible powers ruling the universe or even as gods themselves, and they played an important role in all primitive mythologies. So long as primitive humans were forced to carry on an unremitting struggle with the animals competing with them for the means of existence, they could not assume the place of honor in a world order but had to cede that role to their rivals. Hindus and many savage tribes still hold views of this kind today.

Throughout antiquity and even into the Middle Ages man was obsessed with the idea that animals verged on the supernatural. Pagan imagination invented creatures that were even more terrible than the existing ones, and for many years references to the sphinx, tritons, and centaurs were prominent even in Christian fables. For nearly a thousand years, the *Physiologus* remained the Church's only book on natural history despite the scorn with which many viewed it. The book is essentially a description of 'animal morals in action'. Each creature was the incarnation of a virtue that the true Christian must imitate or a vice that he must shun at all cost.

Throughout the Middle Ages it was believed that animals enjoy special occult powers like those of sorcerers. Roger Bacon believed that the look of the basilisk was fatal, that wolves could make a man speechless, and that dogs could not bark if they were close to a hyena. To a person who had no trouble believing that the goose barnacle is born from a kind of oak tree, nothing should seem impossible. This credulity is no less surprising than that of Pierre Rommel who, in 1680, affirmed that he had seen a cat that had been conceived in the womb of a woman, and that he had known another woman who had given birth to a living goose.

Though such assertions seem ludicrous to us today they are of interest, for they show how confused notions were only a short time ago. It was even believed by

A. McBirney, S. Cook and G. Retallack, *The Philosophy of Zoology Before Darwin: A translated and annotated version of the French text by Edmond Perrier*,

some that under certain mysterious spells animals could give birth to all sorts of other animals or transform itself in the manner of a werewolf. Inanimate material was endowed with the ability to organize itself spontaneously: frogs could be born in mud puddles; old rags left in a box with a little wheat could transform themselves into mice; intestinal worms were believed to be nothing but the metamorphosed humors of our organism, an opinion that, even today, has its supporters.

It was especially difficult to establish a clear demarcation between what was living and what was not. For the ancient philosophers, life was, above all, motion and force. Anything that could move was considered more or less living. Thales of Miletus used the word *soul* for anything that can cause movement. The magnet has a soul no less than man. All forces of nature are living divinities; nature is 'replete with gods.' All things arose from a unique and primitive matter, water, which is itself living and spontaneously fecund.

According to Anaximander, all things stem from an indeterminate substance of infinite mass, the *apeiron*, from which came both heat and cold. These in turn engendered humidity, which generated the primitive soil, which itself, under the action of the sun, produced all animal life beginning with the fishes and extending to all the higher forms.

Anaximenes considered the atmosphere, since it is capable of moving more easily than water, the origin of all things. The atmosphere is infinite, eternal, and living; it is eternally productive.

Anaxagoras was the first to make a distinction between inert matter and a thinking motive force, a kind of coordinating intelligence. Plants and animals were thought to have the same basic faculties. Germs of the former were formed from the atmosphere, the latter came from earth that was fertilized by the ether. He believed that the stars are inert and that one must not attribute to them the qualities of gods.

Thus, most of the philosophers of the ancient world still had a confused concept of organized life. In their view, the stars, the gods, and matter itself are all alive; anything that can move was a form of life and was capable of passing that life on to others. Plants and animals were seen to share certain similarities; they were believed by some to be formed from water, by others from air, and still others from the stars. At the same time, there were attempts to assign a common cause to all forms of life or to attribute them to shared causes. For Thales and Anaximander, all things have been derived from water; Anaximenes and Diogenes preferred to have everything come from air. For Heraclitus, everything is simply a transient form of fire. Xenophon wanted everything to come from air and water, while Empedocles believed that, in conjunction with two universal driving forces, love and hate, all things have been produced from the four elements, earth, water, air, and fire, which, until modern times, have been the basis of all scientific concepts.

There was no place in this philosophy for even the most superficial observations. Imagination was the main means for devising systems, and the sciences that we now recognize did not exist as we know them today. Valid observations were few in number and were mixed with too many fables to serve as the basis of a rational doctrine. Since there was no science, there could be no true understanding of life.

It is worth mentioning some of the attempts to arrive at a more precise explanation of living creatures. Leucippus and Democritus assumed that life consisted of a primitive substance that was separated into an infinity of moving atoms. Bodies differed only in the size, form, and arrangement of their atoms. Those that are finely polished, rounded, more mobile and widely disseminated were the substance making up heat, the soul, and were at the core of intelligence and life.

Anaxagoras accepted the idea of primitive particles, but these infinite numbers of particles are of as many kinds as there are different substances in nature. Bodies are made up of associations of similar particles. It the beginning, all the primitive particles formed a chaotic mixture. They sorted themselves out in some way, under the influence of a coordinating intelligence.

When an animal dies, all its constituent bones, organs, and muscles dissolve and revert to a chaotic mixture. The various invisible parts may then be rearranged and become the integral components of another organism. Thus animals and plants are formed from elements that are permanent and eternal; they come into temporary association when they join to form a living creature, but they then separate again in order to enter new organisms. The elements capable of contributing to these organisms have a constant number, but they are in perpetual circulation, passing from one living creature to another and associating with one another in all possible ways.

The constituent elements of living beings, like those of all other bodies, were said to have existed throughout all eternity and, being indestructible, there seemed to be no essential differences to distinguish living from inanimate material. This view, which was held by Anaxagoras, is not without interest, for it bears a trace of resemblance to the celebrated doctrine of the embodiment of germ cells that we shall come across later in Buffon's hypothesis of molecular assemblages, and in Geoffroy Saint-Hilaire's idea that like elements are attracted to one another. It even bears a resemblance to Darwin's famous theory of pangenesis.[1] Many of these ancient doctrines resemble those of modern philosophers and have turned up repeatedly throughout time. For example, Pythagoras and his disciples claimed that, in addition to the numbers that they believed regulated the world, there were certain fundamental forces that, by balancing one another, governed all of nature: the finite and infinite, the pure and impure, unity and plurality, right and left, masculine and feminine, repose and motion, straight and curved, light and dark, good and bad, the square and circle – ten in all. In this sense, the Greeks were precursors of Schelling and other German natural philosophers who had a similar view of a world in which everything has its opposite. Their views of final causes and

[1] Darwin proposed that minute particles carried in the bloodstream are segregated into gemmules and that, in this way, the mature sexual organs contain a pangene derived from all the animal's organs. The theory has been universally discarded. (Trans. note)

how they were related to these oppositions were quite compatible with the limited knowledge of their times.

This same notion of opposites led Pythagoras to propose the existence of antipodes. Heraclitus held a similar belief and, like the later natural philosophers, believed that our soul is simply an expression of the soul of Earth, of a divine flame. Democritus believed that we have two ways of acquiring knowledge: by our senses and by thought. The senses can be deceptive; only thought can yield true knowledge. Today, Heraclitus and Democritus would be ranked among the members of the 'school of ideas.' For them, as for modern materialists, nothing exists but atoms and empty space. The various ways in which we perceive the external world are the result of motion: we see only change and contrasts, not true objects.

There was little place for observations in these general concepts, even though most attempts to discern the nature of things recognized their importance. For example, Alemaeon of Croton (ca. 520 BC) dissected animals in order to learn more about their internal organs. He compared the white of a bird's egg with the milk with which animals nourish their young, but at the same time he believed that goats breath through their ears. Anaxagoras considered the brain the seat of thought and recognized the manner in which the fetus was nourished, but he claimed that martens gave birth through their mouths and that the ibis and crows copulated with their beaks. Although these two philosophers, and later Polybius, studied the genesis of embryos, we see that they were still capable of reaching faulty conclusions.

Democritus went beyond his predecessors in advancing our understanding of the nature and functions of the organs of animals. Hippocrates made similar contributions, but he focused his attention mainly on human anatomy. He defined a number of common maladies and recognized how they progress. At that time, however, the art of observing, like the art of reasoning, was still in its infancy. As we have just seen, the grossest errors could be mixed with perfectly valid observations. These simple errors had the effect of discrediting the noble intellectual efforts to find insights into the still unexplored realms of science. Science remained inseparable from philosophy, and each advance that philosophers made in the art of reasoning brought further progress in the art of achieving true knowledge. Little by little, intuition took a less exclusive place in speculation, and more rigorous reasoning was used to sort out conflicting ideas. Socrates was the first to give ideas precise definitions and deserves to be honored as the creator of inductive reasoning. Plato showed how much one can achieve by using a method based on a hierarchy of ideas that progress from the particular to the general, but his method was applied mainly to abstract concepts and would require modification before a more rigorous accord could be established between facts and ideas. Little by little, it was seen that well-observed facts are the only reliable source of sound ideas, but a powerful genius was needed to incorporate such principles into ordinary logic without straying from common sense. That genius, the man who would come to represents the foundation of science and scientific method, was Aristotle.

Some of Aristotle's critics have said that his science was drawn in large part from his predecessors, especially Democritus, and that he borrowed much from these earlier philosophers without crediting them. It is the same today; innovative thinkers who propose novel concepts are constantly accused of taking ideas from their predecessors – just as Aristotle was. It is ironic to see the charge of plagiarism aimed at a man who is commonly referred to as the father of philosophy. It is possible, of course, that Aristotle drew on the work of his predecessors – even probable. There is no question that, like any man of great erudition, he would have derived some of his ideas from others. The multitude of facts laid out in his books exceeds by a large margin what he could have acquired from personal experience. Does it then follow that he can be accused of improper use of the good work of others? Accusations of this kind are taken seriously only by those who find satisfaction in making them. Ideas are an essential part of an individual's character, especially when it comes to scientists. This is why genius is so admired. It is also why the efforts of a person whose intelligence approaches the level of genius is received with such impatience by those who recognize their own incapacity to think on this level. It is why anyone who discovers or develops an original idea must expect to be confronted with all sorts of negative reactions, including the usual accusation that he has really done nothing new. In the end, the amount of originality in an idea is of little importance to humanity; it becomes important only when it has been embraced by some powerful mind that is able to show us its significance and say to us: 'Here is a contribution to knowledge, and this is what we can make of it.'[2] The principal value of Aristotle's work was that it summed up all that was known at that time and usually made judicious qualitative distinctions between what was true and what was false. He helped define the limits of human knowledge of his time and pointed the way toward greater certainty in the pursuit of truth. The legacy of knowledge he left to the Middle Ages was so important that, had he not done this, science would have had to go back and start over again from the beginning.

---

[2] The principle Perrier cites here applies especially well to Darwin and the concept of evolution through natural selection which he laid out so lucidly in *The Origin of Species*. (Trans. note)

Chapter II

Aristotle

The earliest notions about the analogies and homologies of organs – Correlative forms – Establishing divisions within the animal kingdom – The concept of species – The principle of continuity – Degrees of perfection of the organs – Possibility that the forms of animals might be transformed.

So much has been written about Aristotle, and the works of this great man are so widely cited, discussed, and interpreted, that readers may be tempted to reproach me for dwelling on a topic that has long since been exhausted by countless earlier authors. But one cannot understand the origins of zoological philosophy without going back to the illustrious tutor of Alexander the Great. Of all the scholars of antiquity, he alone was able to digest far-reaching, rigorous observations and organize his knowledge in a way that permitted him to discern its fundamental consequences. More than one passage of his *Natural History of Animals* could have been written two thousand years later by Cuvier or Geoffroy Saint-Hilaire. The principles of comparative anatomy that Aristotle developed in the first pages of his memorable work are essentially the same as those we recognize today. I need only cite the following lines:

'There are animals whose parts are similar to the corresponding parts of other animals, while there are others that lack these similarities. The parts may resemble each other in having a similar function. For example, the nose, eye, flesh, and bones of a man resemble the nose, eye, flesh, and bones of another man, but they also resemble those of a horse and other animals that we take to be members of the same group... Another kind of resemblance is that of related animals in which the features differ only in the way they are developed. Birds and fishes belong to a single, major group that includes a great number of varieties.

'Members of the same group can usually be distinguished only by qualitative differences, such as color and shape...'

'There are other animals of which it cannot be said that they share similar parts and differ from one to another only in their general form; one can only see an analogy between them. For example, feathers are to a bird what scales are to a fish; the two are comparable. In the same way bones can be compared with arteries, claws with horns, and hands with the claws of the crayfish. In this way, the parts making up the individuals may be similar or they may differ. One must also note their position on the body. Several animals have the same parts, but they are not similarly placed. For example, female breasts may be located on the chest or in the lower region of the abdomen.'

A. McBirney, S. Cook and G. Retallack, *The Philosophy of Zoology Before Darwin: A translated and annotated version of the French text by Edmond Perrier*,

Further on we read:

'As a general rule, most of the parts of animals belonging to separate groups have a different form: some have basic differences and resemble one another only in their general purpose, while others are essentially the same and differ only in outward appearance. They may be found in some animals but not in others.'

Thus the various kinds of resemblances that Geoffroy Saint-Hilaire and his successors would designate as *analogs* or *homologs* were already defined, at least in part, by Aristotle. The philosopher of Stagirus was already conscious of what Cuvier would later call the *correlation of forms;* he cited a great number of such correlations that have since been employed in defining zoological groups. Some of his major ones were:

'All animals have either blood or a liquid, known as lymph, that serves a similar purpose. All animals, whether they have two or four legs or none at all, have blood.[1] All those that have more than four legs have lymph.[2] Animals with blood are larger than animals with lymph, and the size of the latter varies more with climate.

'Animals with hair, as well as cetaceans and selachians, are viviparous, but only the latter have gills; they produce an internal egg.'

The viviparous nature of selachians, which are fish, is quite unlike that of animals covered with hair and differ from the cetaceans, which belong to our class of mammals. Farther along, flying animals are divided into three categories, those that have wings covered with feathers, those that have wings covered only with a layer of skin – the *dermal wings*, - and those that have thin, bare wings made of a membrane. The dermal wings and the wings with feathers belong to animals with blood, and barren wings with a membrane are found only on insects. Insects can have either four or two wings. The coleopterous insects (a term used by Aristotle) whose wings take the form of a case or sheath, have no sting. The insects with four wings have an aculeus [stinger] on their tail; they are the hymenoptera. Insects with two wings have an aculeus on their head. It is clear that Aristotle did not confuse the different characteristics of what he called four-winged, stinging insects with the two-winged insects, for in referring to the latter he wrote: 'a tongue replaces the stinger in the diptera,' and he remarked that insects that have a tongue have no jaw, as though he could see that the tongue that we now call a horn [proboscis] resulted from modifications of the jaws.

He believed that the members of a single group, such as the insects, could be related to one another by clearly defined correlations. Their physiology was also well defined; they were said to consist of separate parts, rings, or segments each of which has a recognizable identity. These parts and segments are what have since been called *body segments* or *zoonites* [somites].

Aristotle's perception was equally apparent in his discussion of mammals. After placing all animals covered with hair among the viviparous, he seems to

---

[1] These are what we now call vertebrates.

[2] Aristotle was thinking mainly of the arthropods and worms.

have feared that the latter might be confused with lizards, which also have four legs, and pointed out that only the tetrapods covered with hair are viviparous.[3] In this way, mammals were clearly distinguished from lizards, to which, as Aristotle showed, snakes bore a strong resemblance, even though they have no legs. All that was needed was a new name to establish the group we now call reptiles.

Other equally remarkable relations were recognized among the four-legged viviparous animals. These tetrapods may or may not have horns. Those with saw-like teeth never have horns, and even animals with no other means of defense have no horns. All tetrapods with horns lack incisors in their upper jaw, and all viviparous tetrapods with horns and no upper incisor teeth possess four stomachs and are able to ruminate. Nothing is lacking in this characterization of ruminating animals. The remarkable correlation of the absence of horns and the presence of canine teeth was precisely expressed,[4] and only in our time has it been explained.

Although Aristotle was familiar with a rather large number of animals, he does not seem to have thought of grouping them in a definite order based on their degree of resemblance. He did not attempt to set up what we would now call a *classification.* Instead, he compared the animals with one another in all possible ways and attempted to reduce the results of his comparisons to very general factors. In doing this, he identified natural similarities that, even today, have a place in modern taxonomy. At the same time, however, comparisons of second-order features led him to make distinctions of less importance. He considered these differences and similarities to be just as important as the primary ones that could have been used if there had been a hierarchy of features that excluded those of secondary importance. Instead of extending to all animals, his comparisons were limited to organisms with similar anatomical structures or, as we would say today, members 'of the same species.'[5]

After having exhausted everything that could be learned from these studies of resemblances, Aristotle went on to consider the differences between various animals. These 'differences in their way of life, their actions, their character, and the parts of their bodies' were all assigned the same level of importance.

Thus Aristotle made definite distinctions between animals that are aquatic and terrestrial, social or solitary, migratory or sedentary, diurnal or nocturnal, and tame or wild. The same animals can, of course, be found in more than one of these diverse categories. In this regard, Aristotle pointed out elsewhere that a given species can include some individuals that are wild and others that are tame. He was

---

[3] Some lizards, such as the *tiliqua* of Australia and *corucia* of the Soloman Islands are viviparous. This is also true of snakes of the Viper family and the North American garter snake, Thamnophis sirtalis. (Trans. note)

[4] This is not true of the fossil record. Several Eocene mammals, such as Uintatheriom, had both. (Trans. note)

[5] Perrier is inconsistent here because he has just shown that Aristotle made comparisons of organisms which we now consider different orders, classes, and families – of reptiles and mammals, of tetrapods or ruminates, and various other mammals. (Trans. note)

not concerned with basing natural groups on what seemed to be fundamental similarities; his purpose, as he saw it, was not to identify and distinguish the different kinds of animals. His book did not deal with zoology but with anatomy and comparative physiology, and he took note of only those distinctive differences that were required for such comparisons. He distinguished animals that have blood from those that do not and divided these into secondary, identifiable groups some of which already had names in ordinary, every-day language. These are what he called the main kinds of animals (γενη μεγιζτα τϖν ζωων). They included birds, fish, shellfish, the molluscs that we now refer to as cephalopods, and even insects. For the latter Aristotle created the new name εντομα, but he rarely made bold proposals of this kind. More often, he used words of ordinary usage, and he expressed regret when he was unable to find such a word for a particular creature he defined. He pointed out, for example, that there was no common name for molluscs and other shell fish and coined a composite word, *ostracoderms* for them; he proposed another composite name, *malacostraces* for lobsters, crabs, and crayfish.[6]

The inadequacies of the common language obviously bothered him. He conceived of mammals as forming a major 'genus,' but popular usage lagged behind him and continued to confuse mammals with the other tetrapods, such as lizards. He did not use the name tetrapod for a natural group, because some tetrapods are viviparous and others are oviparous. Although he chose not to use the word, Aristotle did not propose an alterative. He also recognized natural groups among the viviparous branch of tetrapods but noted that the only group that had been given a name was the λοφουροι, which corresponds to our group of solipeds[7] (horses and donkeys) that are characterized by a cluster of coarse hair at the end of their tail.

It seems that this scarcity of adequate words was the main obstacle preventing Aristotle from arriving at a clear definition of what we now call species. It also made it difficult for him to set up a coordinated system of zoological divisions. The language of his time had only two words to express different degrees of resemblance: ειδοζ, which means *form* or *appearance,* and γενοζ, which is usually translated as *kind* or *type*. A genus normally includes a rather large number of species. Some are large (γενη μεγαλα), and others are very large (γενη μεγιστα), but the forms included in these genera can be divided into additional groups that then become separate genera. When he considered a species independently of its relations to a more extensive group, Aristotle always designates it by the term γενοζ. One can see how this can result in confusion when dealing with a rather complex set of divisions that do not have the same weight and when one uses two words whose meaning changes according to their context within a given division. Nevertheless, if he could not define or even identify a species with a name, Aristotle

[6] Ostracoderms are an extinct type of jawless fish. (Trans. note)

[7] Perissodactyl is the term now preferred for odd-toed ungulates (hoofed mammals) with cheek teeth that enable them to chew plant material. (Trans. note)

certainly recognized their essential attributes; these were the same features we use as criteria today, and they are based on the ability to interbreed.

After defining the types of Lophures (λοφουροι), he included under that name the horse, donkey, mule, pony, and the bardeau.[8] He goes further to say that one should 'also include the dziggetai (half-donkeys) of Syria that bear this name only because of their distinctive appearance. They must be a distinct species *because they couple with their own kind to produce descendents that are fertile.*' Aristotle must have considered animals to be of the same species only if they were descended from the same ancestors, for he used the term *homophiles* to designated animals of similar form. A species is therefore defined by its ability to reproduce with its own kind, exactly as it is today. Unfortunately, Aristotle did not draw the logical conclusion from what is now a widely recognized concept, and he should have been more skeptical of the erroneous tales about certain exotic animals. For example, he did not question the belief that wild forms are more subject to variation in Libya than they are in Greece. He states that 'in very arid regions of Libya, animals assemble at a small number of places to find water, and, while there, the males copulate with females of different species (μη δμοφυλα). The animals resulting from such unions can start a new family line, so long as the two parents are not too different and the periods of gestation of the two species are similar.' A little farther along, he cites the long-held belief that the dogs of India are descended from a female dog and male tiger. When it is a matter of animals in far-off lands, the appeal of something marvelous seems to have obscured the concept of species that Aristotle formed from more conventional observations. He is not surprised that things that happen in Libya differ from what he sees in Greece, because Libya had a reputation 'for producing all sorts of new monsters.' When the strange things he notes from other parts of the world turn up in Greece, Aristotle says that they should just be considered precursors of things to come.

Aristotle's understanding of the different ways in which animals reproduce was too incomplete to permit him to generalize about species. Even though he had precise observations of the lower animals, he could not free himself completely from the opinions that were prevalent at that time. For example, he knew about the eggs of butterflies, fleas, flies and the nest-capsules of the octopus and murex, and yet he declared that these eggs remained sterile. He believed that ostracods, sea anemones, and sponges are born from half-putrefied material that forms on the seafloor and that they differ according to the nature of the sediments. He also thought that butterflies are born from caterpillars but that the latter are formed from green leaves. They were also said to be produced in wood, the excrement of animals, and, under other conditions, from worms that later turn into insects. Is it not surprising that even though the metamorphosis of insects, as well as their reproduction by laying eggs were well known to him, he made no connection between the two? How could such a patient observer remain in doubt about the nature of

---

[8] The product of a female donkey and male horse, the opposite of a mule, sometimes referred to in English as a 'hinny.' (Trans. note)

worms that are only the early larval stage of creatures he knew so well? Aristotle stated that animals that are ordinarily produced from eggs can also be formed spontaneously in the mud of certain swamps.

These ideas were in conflict with the prevailing doctrine that there is continuity in the works of nature, a continuity that philosophers have always searched for and that Aristotle took to be a fundamental law.

In Book VIII he says that 'in nature, the passage from inanimate material into animals takes place little by little and in such a subtle manner that it is impossible to draw a sharp boundary between the two. After inanimate material becomes living plants, the latter still differ from one form to another in the amount of life they possess. Compared to inert bodies, plants appear to be endowed with certain signs of life, but they are still inanimate compared to animals. The passage from plant to animal is even less abrupt than that from inanimate material to plants. In the sea one finds creatures that are difficult to identify as either animals or plants. They adhere to other bodies, and many die if they are detached.' The razor clams and many ostracods, as well as the ascidians, anemones, or sea urchins, and especially the sponges, are among the ambiguous creatures that have some of the characteristics of animals but resemble plants in their apparent immobility.

Studies of animals that are intermediate between aquatic and terrestrial forms led Aristotle to wonder in what fundamental way they differed from one another. This was an opportunity for him to dwell on the kind of philosophical considerations that modern zoologists admire. Animals that live in water are adapted to this environment in several ways: some can breath only in that medium, while others that breathe the open air find their nourishment only in water. Still others require water to breathe but go on to the land in search of food.

Aristotle says that 'the natural features that distinguish these two types of animals can be found in what one might term a kind of discordance. For example, there are males that look like females and females that look like males. A basic difference in a minor part of the body may suffice to bring out important differences in the body as a whole. The effects of castration are proof of this. This operation does not affect more than a small part of the body, but the alteration changes the animal's nature and makes it less different from the other sex. Thus it is reasonable that at the moment when the creature is formed, a seemingly insignificant variation in one of the parts can be a major factor in determining whether the body becomes a male or female. The two types of differences by which I have distinguished aquatic and terrestrial types of animals are based on the disposition of small parts of their bodies.'

Aristotle thought that terrestrial animals could have become aquatic or vice versa, and he attributes the change to fortuitous accidents during their embryonic development. Some of the illustrious naturalists of our time have also postulated that malformations [monstrosities] resulting from accidents of this kind could be an important factor in the diversification of species. According to this passage, Aristotle could be considered what we now call a evolutionist [transformiste], but the question of organic evolution [transformisme] obviously could not be raised at a time when the existence of species had yet to be recognized.

Although his eminently philosophical mind led him through many ramifications of his inquiry, Aristotle failed to recognize a number of important consequences that would become apparent at a later time when our knowledge of animals was more developed. One can see, as Jules Geoffroy points out, how he may have sensed the law of *division of physiological labor* that was developed only in 1827 by H. Milne-Edwards. In Book IV of his *Parts of Animals,* Aristotle states: 'If nothing hinders it, nature always employs two separate organs for two different functions. When that is impossible, it uses the same organ for multiple purposes, but it is better that a single organ not have multiple functions.' Moreover, he did not fail to recognize the 'struggle for survival' that prevails among a great many animals. He says in Book IX that 'animals will always be at war with one another whenever two or more inhabit the same places and compete for the same food. If the amount of food is not sufficient for both, they will fight one another, even if they belong to the same species.' In his *Physics*, Book II, Chapter VIII, Aristotle even wondered whether this struggle could result in the extinction of forms that were insufficiently adapted to the conditions under which they were living and to the survival of others that were better adapted.[9] He rejected the idea that supernatural forces governed the natural world; if they played a role, it was only in exceptional circumstances. He considered the resources of nature great enough to make the destruction of one of its works impossible. All animals are not necessarily in conflict; there are some that are quite compatible with one another. These are only a few of the brilliant passages in his *Natural History of Animals* in which he described the behavior of the animals he studied and he showed that he was just as capable as an observer as he was as an anatomist.

In summary, this immense work, the general features of which I can only sketch, can best be referred to as 'Zoological Philosophy.' Aristotle assembled facts only so he could use them to arrive at general laws, and his penetrating mind discerned in them a bountiful array of general relationships. Some of these are of only passing interest, but several of those he laid out in his *Natural History of Animals* have been firmly incorporated into science exactly as Aristotle formulated them. What may be most marvelous of all is that, from the outset, Aristotle made use of all the different approaches by which the animal kingdom can and must be studied: comparative anatomy, physiology, embryology, animal behavior, geographical distribution, and the inter-relationships that these factors have with one another. All of these were subjects of his studies, and his research left us one of the richest treasures of knowledge that the human mind has ever possessed.

---

[9] Perrier's interpretation differs somewhat from that of Henry Osborn who, in his book, *From the Greeks to Darwin* (1922), points out (p. 49–57) that Aristotle rejected Empedocles' explanation of adaptive development through the survival of the fittest. Instead, he favored a form of Intelligent Design that brings gradual perfection over time. 'Had he accepted Empedocles' hypothesis,' Osborn remarks, 'he would have been the literal prophet of Darwinism.' (Trans. note)

Chapter III

The Roman Period

> Lucretius: Formation of the earliest organisms; the struggle for life. – Pliny: attributed marvelous things to animals; the nature and origins of marine monsters; notions about anatomy. –Elien, Oppien, and Galen: Progress in anatomy; correlations between the external form of animals and their organization and behavior.

It would seem that after Aristotle set science on its true path, it could only to go forward, and one would think there would have been a marvelous flourishing of science after the appearance of that great man's work. Unfortunately, the political divisions, wars, and invasions that prevailed at that time prevented the continuation of Aristotle's work in the East where it had begun. Aristotle was soon forgotten, and, most surprising of all, when his work reappeared, it did not set off a scientific revolution but actually became an obstacle to progress. His great work inspired such admiration that it was accepted without question or proper understanding. The master's opinions became dogma; the literal sense of every sentence he wrote was endlessly discussed, while the great principles he left us were totally ignored. When questions arose, scholars failed to follow his example and never looked to nature for the evidence needed to address them. Instead, they carried on endless arguments about the meaning of his words and only added to the confusion. During the Middle Ages, Aristotle was held to be a kind of pagan Moses whose words were as infallible as those of the holy book, and a major effort would be required before science would be able to recover its open, independent character.

As antiquity came to a close, Rome could have taken up the role of Greece and transmitted to the West an echo of the brilliant philosophical essays of that privileged nation, but Rome was too engrossed in the life of the forum and too preoccupied with multiplying and extending its conquests for its philosophers to find the leisure needed to observe nature. There were, however, a few individuals with astonishingly penetrating minds. One of these was Lucretius.[1] This man's magnificent poem contains a number of prophetic views that would later be confirmed by modern science. For him, the Earth was the mother of all living things. Like all organisms, it has had a period of productivity during which it produced most of the plants and animals, but that stage soon passed, and it has now entered a period of relative sterility.

---

[1] Lucretius (ca. 99 – 55 BC) a Roman poet and philosopher, left only one known work *De rerum natura (On the Nature of Things)*. (Trans. note)

'In the beginning, Earth covered her hills with a fresh mantle of vegetation and adorned all the verdant fields with flowers. A magnificent struggle developed between the different trees, each striving to send its branches higher into the skies. Just as a coat of fur developed on four-legged animals and feathers covered the bodies of birds, the youthful earth was soon covered with herbs and shrubby trees. By various means, it later created the innumerable varieties of mortal creatures. Animals did not simply fall from the skies, and plants could not have emerged full blown from the depths of the sea. So let us give Earth its well-deserved name, 'mother,' in recognition of all the creatures that have been nurtured at its breast. Even today, many living creatures are formed in the ground with the aid of rain and the warmth of the sun… In the earliest centuries, many races of animals were doomed to disappear because they were unable to reproduce and perpetuate themselves. All the forms of life we see around us are protected from destruction by their innate craftiness, strength, or agility. Many benefit from their usefulness to humans and survive only because we defend them. Lions and other cruel, ferocious beasts are protected by their strength, the fox by its cunning, the elk by its ability to run swiftly. The faithful and vigilant members of the dog family, the flocks of wool-bearing sheep, and many animals with antlers have long been under man's protection… But why would we want to protect useless animals that nature has not endowed with the qualities they need for their independent survival? Doomed by their fatal weaknesses, these creatures served as prey for their rivals and with time vanished completely.'[2]

Could there be a more brilliant statement of the doctrine of the struggle for life, the extinction of species that are insufficiently endowed for survival, and the *natural selection* that is its consequence? Lucretius believed that when living creatures were produced in nature the simplest forms were the first to immerge and any that were imperfect were destined to disappear and be replaced by new ones that continued to appear. Is it not astonishing that he stopped at this point without seeing that the appearance of the earliest simple species could lead to the more complex species that followed them? The poet did not understand the true nature of fossils, nor did he recognize the powerful destructive forces against which all forms of life struggle to survive. He thought that those forces were most important during Earth's early period of exuberant productivity when the malformations [monstrosities] it produced would disappear almost immediately. Today, he argued, this natural process no longer has an effect. Although he employed words like *corda* or *sæcla* to designate species in terms that implied a continuous series, he does not seem to have believed that an intermediary was needed between the common mother and its first children. In short, it does not seem to have occurred to him that the forms living today might not be immutable. And, unlike Aristotle, he did not suspect that they might vary.

---

[2] Lucretius, *De rerum natura*, book V, verses 781 to 875. [The text given here is from Perrier's condensed French translation of the Latin original.]

Lucretius did not trouble himself with factual details. In this respect, he was quite different from Pliny, whom we usually think of as the greatest naturalist of the ancient world after Aristotle. Earlier philosophers had already constructed a system for explaining the world. We can use an expression like the one Buffon used in referring to himself: Aristotle assembled facts in order to derive ideas from them, whereas Pliny confined himself to assembling facts. He did not take them solely from nature but from wherever he could find them, and in this way he produced a vast compilation in which valid observations were indiscriminately mixed with all sorts of fables dating from ancient mythology down to his own time. He made no attempt to judge these stories critically.

On every page of Pliny's *Natural History* one finds the idea that animals are intimately aware of even the most obscure resources of nature: they know a host of medical remedies, they know how to observe the heavens,[3] forecast the winds, rain, and storms, and they can make all sorts of predictions. When a house is in danger of being ruined, rats leave it and spiders fall from their webs. Birds forecast even the most minor events of human life. The people of Thrace considered the fox an excellent source of guidance. The hyena was a true magician. The flesh of the bear continues to grow after being cooked. The wind could make certain oaths come true. Pliny did not find any of this in the least bit surprising, for he believed that the germs of all things fall from the sky and that this explained how the sea could produce huge animals and extraordinary monsters. Germs accumulate in the immensity of the sea and provide abundant nourishment to its inhabitants. Mixing with one another indiscriminately, they give birth to all sorts of strange beings that may simulate animals or inanimate objects that one observes on land. This can result in odd assemblages, such as the tiny sea horse, a fish with the head of a horse.

A few quite valid observations can be found within this singular doctrine. For example, many authors maintained that fish cannot possibly breathe because they have no lungs, but Pliny says: 'I must admit that I cannot accept this view, because under certain natural conditions animals can have respiratory organs other than lungs, just as many animals have a bodily fluid other than blood. Who can doubt that air can go into water when we see that it can come out of it?'

In dealing with marine animals, Pliny did not confine himself to fish; he also described the octopus and several molluscs, and he stressed the close association of mussels and razor clams that Aristotle had already noted. He wondered whether the sea urchins, jellyfish, and sponges might not share certain features of both plants and animals, but he was less perceptive than Aristotle when he ranked whales with fish and bats with birds, and he showed that he considered similarities and differences in the structures of animals more important than similarities and differences in their way of living.

[3] Book VIII, chapter XLII, §27 and 28.

Bees held the place of honor among the small number of insects that Pliny described. Then came the wasps, hornets, spiders, scorpions, crickets, scarabs (what Aristotle called beetles), grasshoppers, ants, and, along with all these articulated insects he included the geckos, which are, of course, reptiles. Pliny believed in the spontaneous generation of many of these creatures. The tiny droplets of dew that condense on the leaves of cabbages can produce a caterpillar that then becomes a chrysalis and finally a butterfly. Moths are born from dust, and flies and meal-moths are produced by fire.

The ritual sacrifices that were thought to cause the oracle to make prognostications of future events provided the means by which Romans developed a rather precise knowledge of the physiology of mammals. Pliny devoted an important part of his *History of Animals* to descriptions of their main internal organs and their functions. Some of his notions are quite correct, but they are mixed with countless fables. He refers to birds with two hearts and others that had none at all. He said that the number of lobes on the livers of rats was always equal to the number of lunar days. Apart from the principal internal organs, his knowledge of anatomy was very limited. The veins, arteries, nerves, and tendons, although crudely distinguished, are confused with one another, and Pliny had no understanding of their functions. He said that birds have neither veins nor arteries, that claws and fingernails are the terminations of nerves, and so on.

Despite these short-comings, Pliny is the only Latin author whom one can reasonably call a naturalist. Elien was simply a compiler, and if the work of Oppien showed that the Romans possessed interesting information on the behavior of animals, the titles of his poems: *Cynégétiques* [Hunting with dogs], *Halieutiques* [On Hunting], and *Ixeutiques* [Catching Birds], show that they had been composed with other purposes in mind.

A single great figure appeared before the Roman Empire began its final decline: Galen.[4] Galen was primarily a medical doctor, but he possessed a remarkable philosophical mind and laid out an enlightened program of scientific education for which he wrote a series of treatises that dealt with everything from the art of speaking and reasoning to medicine. An excellent observer, he advocated a close alliance between reasoning and observation.

Since he could not dissect human bodies, he studied monkeys, primarily the rhesus. He laid out for his reader informal ways of studying the skeleton, to which he was the first to apply that name. He urged his readers to explore old collapsed tombs, where one could find the desiccated bodies of brigands, and suggested going to Alexandria where skeletons were available for study. He advocated systematic studies of bones, muscles, arteries, veins, nerves, and the intestines, and he deserves credit for distinguishing the nerves from tendons and showing that all the former lead to the brain or the spinal cord. Galen identified their functions through well-designed experiments. He saw the presence or absence of nerves as a basic

[4] Galen's medical theories prevailed for more than a thousand years, especially in the Islamic world. He was a strong believer in experimentation and empiricism. (Trans. note)

distinction between plants and animals. He knew that veins and arteries both carried blood, and his observations on the functions of organs were a major improvement over what was previously taught.

Because it was impossible to dissect human bodies in a methodical way, Galen had to carry out studies of other types of animal life, and this led him to make a number of interesting comparisons. He even came to recognize that all the living creatures he had studied shared a remarkably uniform structure. He wrote that 'what you observe in the nutritional organs may seem incredible, but if you look closely your doubts will be dispelled. You will marvel at *the way in which these parts demonstrate that a single artist has constructed all the animals and has designed their organs in a way that makes them appropriate for their intended use.*' Thus, like others, Galen saw unity in the diversity of life.

He obviously believed in final causes but concluded from the relationships he saw between organs and their functions that there is also a relationship between external forms and the internal organization of the body and between the behavior of animals and the structure of their bodies. 'The parts that serve similar functions and have the same external form must necessarily have the same internal structure. All animals that have the same physical behavior and external forms also have the same physiological organization. Nature, in effect, has given to each animal a body that is suited for the faculties of its soul, and this is why each creature, from its birth, uses its organs as if it were following the instructions of a master. I have never dissected small creatures, such as ants, mosquitoes, or fleas, but I have dissected animals that carry them, such as weasels, rats, snakes, and many types of birds and fish, and I have become convinced that the same intelligent being produced them all. They all have bodies that are adapted to their behavior. *One can deduce the internal structure of animals from a superficial examination, even without dissecting them, and it is even easier to see this if one can also study how the parts of the body function.'*

This is essentially the principle of adaptation to conditions that Cuvier would later propose in almost the same terms. Cuvier postulated that the rules that Galen clearly perceived for correlating the external form of an animal with its internal structure had been established by higher powers, and he followed these same rules when he used the relationship between the parts of animals and their internal organs to reconstructed fossil animals. Just as Aristotle's wisdom has been attributed in large part to the work of his predecessors, one could with equal reason give Galen credit for much of Cuvier's work. And as we have just seen, one could even say that it was he who inspired Geoffroy Saint-Hilaire's principle of a unified plan for the bodies of all animals.

Chapter IV

The Middle Ages and Renaissance

Arabian medical doctors. – Alchemists. – Albert the Great. – The first great voyages. – Renaissance of anatomy. – Belon, Rondelet. – Francis Bacon. – Progress in physiology and anatomy. – The first micrographs. – Prejudices that reigned until the XVIth century.

Galen was the last great philosopher to lend intellectual vitality to the Roman world at a time when the empire was in a state of general decline. Soon the barbarians would be surging into all parts of the crumbling Roman civilization. Paganism was in retreat, and the establishment of Christianity absorbed all the intellectual efforts of those who still had time for such things and were not totally preoccupied with warfare. All scientific culture was disappearing in the western parts of the empire, and it was only in the east that the wealth of knowledge amassed during antiquity was preserved by a different race of men who were caught up in the exuberance of a new culture. Throughout the Middle Ages, it was the Arabs who were the dominant scientists. Starting in the IXth century, the medical sciences began to flourish, and Hippocrates and Aristotle were translated into vernacular languages. The most famous names from this extraordinary period are El Kindi (860), El Dehadidh - the author of a natural history of animals, Abou Hanifa - a skilled botanist, and Ibn Wahchjid. They blended magic into science and metaphysics. Rhazes (850–923), Avicenne, Avenzoar (1070–1161) and his student, Averrhoes (1120–1198) made the medical profession more respectable, but they were inclined to indulge in speculation at the expense of sound observations. As it was for most scholars at that time, philosophy was more important than knowledge, and, if they helped preserve the scientific tradition of the ancients, they did so without contributing much that was new to anatomy, physiology, or medical diagnosis. They had, however, an extensive knowledge of the medicinal properties of plants, and we owe to them the introduction into therapeutic medicine of a great number of medicines. Kazwyny (1283), Ibn el Doreihim, El Demiri (who lived in the XIVth century), El Calcachendi (1418), El Schebi, and El Sojuti (1445) wrote remarkable treatises on animals and described their principal traits. El Demiri wrote a kind of dictionary of natural history that included descriptions of 931 animals.

It was to these Arabian doctors and naturalists that the European scholars of the Middle Ages now turned their attention. It is largely to their influence that one must attribute the strange mixture of astrology and alchemy with true science that

A. McBirney, S. Cook and G. Retallack, *The Philosophy of Zoology Before Darwin: A translated and annotated version of the French text by Edmond Perrier*,

one sees throughout this period. It created a melange which even the greatest minds of the time were unable to ignore; all distinctions between science and sorcery had disappeared. Even Roger Bacon (1214–1292), while protesting the uselessness of magic, was taken in by the fascination of alchemy. A man of broad interests, he was an ingenious researcher and skilled experimentalist. Certain passages of his *Opus majus* would lead one to suspect that he foresaw many wonderful modern inventions; he even seems to have understood the art of fabricating gunpowder. He ranks among the men who did the most to guide scholars toward objective observations of nature.

The studies of this period proliferated into a wide range of scientific fields. The practice of medicine merged with various philosophical and even theological topics, including the search for the philosopher's stone and the transmutation of metals. Most naturalists confined themselves to writing theological commentaries on Aristotle's works, and if they contributed observations of their own, they usually followed a conception of nature in which nothing seemed impossible. This way of thinking made it difficult to separate outward appearances from reality, and one almost wishes that these scholars had confined their laborious writing to discussions of the ancient texts. This is especially true of the alchemists, even though many of them earned well-deserved reputations for their writing and contributions to other fields: Arnaud de Villeneuve (1238–1314), who discovered alcohol, Raymond Lulle (1235–1315) to whom we owe nitric acid or *aqua-forte,* and Albertus Magnus (1153–1280), a Dominican who became bishop of Regensburg but abandoned that honored post in order to devote himself to scientific studies. He exercised a strong influence through the numerous works on alchemy in which his thinking had a strong theological bias. One of his disciples was Saint Thomas Aquinas (1227–1274) to whom Pic de la Mirandole dedicated a work on alchemy and whom the Catholic Church still places in the highest ranks of men of science.

During the XIIIth century a number of exploratory voyages, such as those of Guillaume Rubruquis and Marco Polo, brought back new knowledge of eastern Asia. Marco Polo was the first to reach China and Japan, but because the account of his travels was not in accord with what Aristotle had written, it was long considered a product of Marco Polo's imagination.

Despite the invention of printing in 1434 and the great voyages of Christopher Columbus who, in 1492, discovered America, the scientific errors of the XIIIth and XIVth centuries continued to prevail well into the XVth century. It was not until the XVIth century that new light finally began to illuminate the minds of scholars, and important scientific research was undertaken. André Vésale (1514–1564) regenerated anatomy; Fallope, Eustache, Spiegel, Ingrassias, Botal, and Varole, had their names attached to some of the organs and structural features of the human anatomy that they studied. The research of Fabrizio d'Aquapendente (1537–1619) and that of two outstanding botanists, Collombo and Césalpin, paved the way for the discovery of the circulation of blood, for which Césalpin provided an excellent general description. The pulmonary system was clearly recognized by the unfortunate Michel Servet (1509–1555), whom Calvin caused to be burned as a heretic in Geneva. This was also the period in which the famous surgeon of Henri II, Ambroise Paré (1517–1590) made the first comparative study of the skeletons of birds and mammals.

In addition to these advances in anatomy, there was a similar renaissance of botany and zoology. Jean and Gaspard Bauhin, the first of whom died in 1613 and the second in 1624, published important works on plants, even while pursuing their innovative work in medicine. Pierre Belon, born in 1518 and assassinated in the Bois de Boulogne in 1564, wrote a *Histoire naturelle des animaux marins* and a *Histoire des oiseaux*. When he compared the organs of various animals he opened the way for comparative anatomy. At the opening of his work on ornithology he presented a drawing of a bird's skeleton beside that of a human and designated with the same letters the parts that seemed to him to correspond in the two skeletons. Around the same time, Rondelet (1507–1566) produced a very beautiful *Histoire universelle des poissons*, in which one finds an early attempt to set up a system of classification. But among the naturalists of that century, the persons who displayed the most remarkable knowledge were Conrad Gessner (1516–1565) and Aldrovandi (1527–1605). Gessner published various philosophical and scientific works, as well as a four-volume *Histoire naturelle des animaux* and several descriptive botanical works in which he set up, on the basis of reproductive systems, the first scientific classification of plants. He also looked into crystals and fossils and postulated that the latter could well be the remains of living creatures. Aldrovandi was the author of a vast natural history in which he dealt with all three realms of nature. His work was supported in part by the Bologne senate.

One of the notable achievements of the great artist Bernard de Palissy (1500–1589) was to champion the idea that most fossils were the remains of marine life and that this meant that the seas had previously covered vast extents of the continents, an opinion already put forward at the beginning of the century by Leonardo da Vinci.

Little by little, trust in observations, experiments, and reasoning was substituted for blind faith in the authorities. Scholars finally gave up their endless discussions of the opinions of classical masters, but they had an unfortunate influence on the spirit of Christianity when they began to express an ill-concealed disdain for Christian dogma. It eventually became apparent how sterile and vain these disputes really were, and there were pleas for a return to the kinds of observations of nature that Aristotle had taught. A number of investigators advocated following Aristotle's example but without regard for classical authorities. A few hardy individuals like Argentier proclaimed their exclusive dedication to reason and prepared the way for Francis Bacon (1561–1626), whose *Instauratio magna* finally restored the scientific methods that had been ignored since the time of Aristotle.

Bacon declared that scientists must put their trust primarily in what they learn from experiments, and he even extended the experimental method to studies of the origin of life. In his *New Atlantis* he provided a fresh impetus to the natural sciences and recommended *testing the metamorphoses of organs and using the variations in species to study how they reproduce and diversify*. This is the first scientific expression of the idea that the forms of plants and animals are not immutable and finite in number; he believed that all life arrived at its present state through a series of slow and gradual modifications. By the time he died, this illustrious philosopher was able to understand the importance of one of the most wonderful discoveries to come from use of the experimental method, namely the circulation of blood announced in 1619 by Harvey, the doctor of James I and Charles I. Harvey was a student of Fabrizio d'Aquapendente whom he had assisted in studies of the coronary valves and veins. His discovery opened an entirely new approach to anatomical research.

Aselli drew attention to the chylous vessels that Pecquet had shown are designed to extract assimilable material from the entrails and transport it into the thoracic duct through which it is introduced into the blood system. Rudbeck and Bartholin both claimed credit for the discovery of the lymphatic vessels; Wirsung identified the pancreatic canal; Bartholin and Sténon carried out a study of the salivary glands; Wepfer, Schneider, Willis, and Vieussens extended our knowledge of the brain and helped define its role. Ruysch introduced a procedure in which colored liquids are injected into the blood vessels and used it to make great progress in understanding the vascular system.

Around this same time, another method of investigation that was used in studies of organisms proved even more fruitful. Almost simultaneously, Malpighi, a professor of medicine at Bologne (1628–1694), Leuwenhoek (1632–1723) in Delft, and Swammerdamm (1637–1680) in Amsterdam introduced the use of magnifying lenses into studies of nature. Malpighi recognized a great number of peculiarities in the structure of human organs, discovered the windpipes of insects, and studied the development of the chicken. We are indebted to Leuwenhoek for having drawn the attention of naturalists to infusoria and for having contributed to the discovery of sperm. He also seems to have recognized the asexual reproduction of the plant louse, as was later verified by Bonnet in Geneva, and his observations on the generation of polyps by budding were overlooked until the research of Trembley drew attention to them. Swammerdamm, who published a great part of his work under the title *Biblia naturæ,* is famous chiefly for his research on the metamorphosis of insects.

It was during this same period that several important questions became topics of lively discussion throughout much of the scientific world. Rédi used well-designed experiments to discredit the hypothesis of spontaneous generation. He conceded that this form of generation might be possible in the case of worms found inside fruit and in the intestines of humans and animals, but he argued that it was the vital forces themselves, the embryonic souls, *souls of vegetation*, that engendered these worms. Near the end of his book on *Optics,* Newton pointed out the uniform structure in animals, a principle to which Geoffroy Saint-Hilaire would devote his entire scientific career. Pascal went beyond Bacon and proposed that *animals were originally crude, ill-defined individuals, and their physical make-up was determined by the prevailing conditions in which they lived.*[1] Sylvius Leboë of Leiden maintained that all the processes that take place in the vital organs are analogous to reactions carried out in chemical retorts in a laboratory, while Vallisneri tried to explain the generation of life as the result of preformation [emboitement] of germs, a doctrine of which Cuvier was one of the last partisans. Swammerdamm laid out the basis for the theory that animals develop by forming successive parts by *epigenesis*, but the thinking at that time was still far from being able to recognize the significance of these discoveries.

In 1595, Frey, a pastor in Schweinfurt, considered animals to be 'preceptors' that have been given to us by God. Wolfgang Franz, in his *Sacred History of the Animals* (1612), proposed a rather ingenious classification of animal life that included dragons that have three rows of teeth in each jaw and added with ineffable serenity: 'The most important dragon is the devil.' P. Kircher, a distinguished physician, studied the animals that Noah is said to have taken aboard the ark and included among them sirens and griffons - and this was in 1675! These were not true scientists but religious writers. They illustrate the extent of prejudices that still confronted any sort of scientific discovery!

[1] Geoffroy Saint-Hilaire attributed this sentence to Pascal, and its context resembles that of writings by the author of the *Provincals*, but Isidore Geoffroy Saint-Hilaire and Jules Soury have searched for it and have been unable to locate its source. I have been equally unsuccessful, and its authenticity remains doubtful.

Chapter V

Evolution of the Concept of Species

> The great descriptive works: Wotton, Gessner, Aldrovandi. – Ray: definition of species. – First attempts at nomenclature. – Linnaeus: the fixity of species; binomial nomenclature.

Descriptive zoology was now making real progress. In 1552, Wotton wrote an essay in which he used the work of Aristotle to deal with the animal kingdom in a systematic way. The same year, Conrad Gessner's *Histoire des Animaux* compiled all that was known about the animals living at that time and made comparative studies much easier by adopting methodical descriptions. Starting in 1599, Aldrovandi published a series of important works on animals and set up a rigorous classification, part of which was borrowed from Wotton. Mythical animals like harpies and griffons were still mixed with the real animals, and even though he repeated stories such as the one about a goose that was born from acorns, his work was recognized as an important step forward. Similarly, Jonston drew on other works on natural history to compose his *Théatre Universel des Animaux* and here again the theme was the same: the animals were described according to their normal habitats, their sources of nourishment, and behavior.

But more and more varieties were now recognized, and it was becoming increasingly difficult to sort them out by means of the long, confused descriptions that had been used until that time. In 1661, Sperling was the first to conceive of the idea of defining them by means of simple factors that he called *precepts*. The groups of animals to which these rules were to be applied were clearly defined in the minds of zoologists but could not yet be designated by particular names. Just as Aristotle did much earlier, Sperling used the terms genus and species indiscriminately to designate vaguely defined groups. It was said, for example, that the bird species included a great number of other animals and that the mammals were divided into several genera. Despite Rédi's efforts to show that spontaneous generation of insects was simply nonsense, it was generally accepted that some animals occasionally gave birth to other quite different types of animals, and that many creatures can be born from dew drops, rotting material, or slime.

The need for more precise definitions was becoming increasingly apparent. John Ray took the same approach we follow today in order to determine exactly what we mean by the word *Species,* and he established a set of criteria that could be widely accepted. He used the name species only for the most restricted groups to which that name had previously been applied, and all species that share certain

A. McBirney, S. Cook and G. Retallack, *The Philosophy of Zoology Before Darwin: A translated and annotated version of the French text by Edmond Perrier*,

characteristics would belong to what he called a genus. Genera could now be divided into species, and the species became separate, subsidiary units. His definition was based entirely on an integration of commonly observed facts. The plants and animals we know best have originated only from plants and animals that they closely resemble. They alone are genetically linked in what would now be called species. Although Aristotle did not use that word, the general idea was already apparent in his work. The concept was expressed less precisely, however, because Aristotle rarely spoke of it except in reference to difficulties he saw in the relationships of certain animals. Ray, on the other hand, expressly stated that 'the offspring of distinct forms always have the same outward appearance as their parents. One species is never born from another one.' It seems that, in proposing clearer criteria for defining species, Ray believed that these forms were immutable, but he was not yet ready to state this expressly. He saw that there can be notable sexual differences among animals of the same species, but he also noted that his 'characterization of the species is not absolutely infallible. Experiments show that some seeds can degenerate and that in exceptional cases a parent plant can give birth to different species and in this way lead to the transmutation of species.' These reservations would soon disappear, however.

The broad scope of Ray's studies embraced botany and almost all branches of zoology which he studied either alone or in cooperation with his friend Willoughby; when Willoughby died prematurely, Ray published his work. Little by little the growing collections of animals from all parts of the world forced naturalists to focus their studies on specific collections, which they described in meticulous detail just as one might now describe a collection of curiosities. This led to books like Seba's *Thesaurus,* Rumphius' work on the rarities of Amboine (1705), Pétiver's *Gazophylacium naturae et artis* (1705), and other publications of a similar nature.

One could, of course, limit one's descriptions to a particular category of animals that have some resemblance to one another, but these categories were set up on the basis of certain natural associations of animals that had already been recognized. Martin Lister, for example, worked on common shellfish, Breyn on sea urchins, Linck on starfish, etc. The monographs coming out of these various studies never generated broad ideas, but they showed the importance of studying living forms. These forms were clearly defined and often carefully illustrated as, for example, in Linck's work on starfish that was published in 1733.

Among the groups that were studied in this way, those that shared the greatest resemblances were the genera that constituted a secondary division within a more extensive group that the author had selected as the subject of his study, but these had not yet been assigned a proper nomenclature to indicate how closely they were related. In the works of Breyn and Linck, each genus was given a particular name, and each species was distinguished from others of the same genus by one or two modifiers attached to their generic name. This system of nomenclature resembled the one used in civic affairs and found increasing use in the language of zoology. At first, the nomenclature tended to be rather informal. Several modifiers were used to designate a single species, but it was eventually recognized by Linnaeus that it was necessity to formulate a strict set of linguistic rules. In 1749 he published his now famous *Pan suecica* in which he informally designated the common Scandinavian species by names and unique modifiers, and in 1751, in his *Philosophie Botanique,* he pointed out the advantages of this method of nomenclature. In 1753, in his *Species plantarum,* he was the first to apply a nomenclature of this kind to plants, and in the 12th edition of his *Systema naturæ* (1799) he extended it to animals as well. This method of designating species, which we now call the *binomial nomenclature*, has since been widely adopted by all naturalists.

By a phenomenon that was in some ways the reverse of that which prevented Aristotle from grasping the notion of species, many groups were clearly defined and designated by definite, easily remembered names, but these groupings were artificial and soon proved to be unrealistic. During the period that was now opening, naturalists lost sight of the need to use the distinctive identities of species to assign them to a particular group; instead, the criteria they were now using became increasingly abstract. Many naturalists seemed to think that the main goal of science was to recognize all the living forms and catalog them as completely as possible. Klein was probably the most accomplished practitioner of this doctrine. His works served the sole purpose of creating a catalog in which the animals are easily identified, and it was his view that to achieve this one must have a classification that is based exclusively on the external forms of the animals. It is true that these easily recognized characteristics are the most practical way to determine the place of a given animal in this sort of inventory of the animal kingdom, but it is equally important to describe not only the physical nature of these characteristics but also the purposes they serve, so that these factors can be used as what we might call *criteria for classification*. Artifacts, such as the binomial tables used by botanists, are eminently useful when one is simply looking for a name to apply to a particular creature. Klein maintained that a naturalist who wants to find the name of an animal should not have to open its mouth and count its teeth. Instead, he preferred to cite the descriptions provided by naturalists. It is regrettable that, even today, some of our methods of classification are not based on such principles.

It was Linnaeus who had the honor of constraining the influence of Klein's ideas. He affirmed that natural history should have a higher goal than that to which simplistic nomenclatures threatened to confine it. There is harmony in nature, and any naturalist worthy of the name should strive to understand it. He did not deny that a developing science had to make use of more or less artificial procedures to establish a simple inventory of all forms of life so that they could be easily identified and distinguished from newly identified forms. And it is true that he owed his own brilliant reputation to the invention and general use of ingenious procedures of this kind, but he considered these procedures, which he called *systems,* only a temporary concession to the needs of nomenclature, and he did not claim that they had any scientific basis. To him, everything in nature appeared to be rigorously ordered. He was convinced that all creatures are related in a logical fashion, much as our thoughts are linked to one another in an uninterrupted chain.

He was also in accord with the aphorism that Leibnitz had stated: *Natura non facit saltum* – Nature never moves by leaps. Each species in the long series of living forms should fit neatly between two others. Scientists should strive to place species in this kind of order, for only then can they be confident that their system of classification is definitive. Such a system would not necessarily be unique, and it should be referred to as a *natural method.* Linnaeus thought that this could be achieved by setting up a series of procedures of this kind and then perfecting them by successive refinements, so that they would gradually merge into a more and more definitive system. Thus, each of these tentative systems resembles a theory that initially offers only approximate explanations for the phenomena it is meant to relate to one another, but with time, progressive improvements made it possible to give the relationships firmer cohesion.

This method would portray nature as a faithful manifestation of the plan of the creator. It should take into account all known facts that have been learned from studies of animals. Consideration should be given, not only to their external forms but also their anatomical structure, their faculties, and their way of living, so that species can be placed in their natural order. Although Linnaeus limited himself to setting up what he called a *system of nature,* he introduced, so far as it was possible at the time, the notion that the animal kingdom has an orderly structure. The new approach that he opened would later be followed by Cuvier.

The illustrious Swedish scientist made another contribution that was equally important to the future development of zoological philosophy.

The first thing that was needed to achieve his highest goal was to introduce more precision into science – something that was sadly lacking in the work of that time. Everything he spoke of was very carefully defined. Other naturalists would have seen little need to define the meaning of animal, vegetable, or mineral; it was obvious from everyday observations what these terms meant, but Linnaeus wanted to be more specific. He stated that:

Mineralia grow
Vegetalia grow and live
Animalia grow, live, and have senses.

When the three realms are characterized in this way, their characteristics are seen to grade from one to the other in a logical progression. Individual forms are defined with no less clarity: 'We take into account all the species that have come in pairs from the hands of the Creator.'

The divisions he proposed were probably a bit too precise, for they were meant to deal in a concise manner with a host of questions that it might have been wiser not to try to resolve too quickly. Linnaeus appears to have believed that all animals have come in pairs from divine hands, and that all animal species we observe today have descended from these pairs through an unbroken series of generations. He assumed that none of the natural families that originated in this way has become extinct and that they have never mixed with one another. They have not been perfected, degraded, or modified in any way. This belief could not be supported by even the simplest observations or experiments; it was merely the logical implication of a definition of species that had no scientific basis. It is evident that Linnaeus was inspired by the account of creation in the book of Genesis. We are presented, not with rigorously determined facts, but with a religious belief. What he had just introduced into science was actually a dogma. It is true that he did not attach excessive importance to this belief, and it did not prevent him from carrying on sound research into the variations to which living creatures are susceptible. His work led him to conclude that primitive species of plants were far from numerous and that their numbers have increased over time by additions of what are now well-established species. In defining species as he did, Linnaeus filled the need for a sharper expression of a scientific notion that was still vague in the minds of most of his readers. Henceforth, his students and successors would use the most meaningful elements in this definition to derive a principle that made the invariability of species the keystone of zoology. Linnaeus had said that 'every species is exactly intermediate between two others' and that 'Nature does not move by leaps.' In his view, these two propositions implied a deep sense of continuity in the animal kingdom, just as he believed there was in the plant kingdom. This tended to temper the rigor of his definitions; his successors would soon show that there are in fact discontinuities.

The Linnaean school has often been accused of being an impediment to any studies that could clarify the origins and possible modifications of living creatures. This reproach is not well founded. Whatever the spirit in which they are made, precise observations always provide factual information that ultimately leads us to the truth. Linnaeus endowed natural history with a precision that was unknown until his time, and even though he believed that living forms are immutable and limited in number, he made it possible for naturalists to agree on the number and characteristics of distinct forms that are clearly distinguished from one another. If these forms could in fact vary, descriptions of supposedly new species continue to add to their endless numbers, and with time transitions from one to another were identified. Some of the transitions may be abrupt, others gradational, and some of the intermediate forms are alive today, while others have long since disappeared. Need I say what this has done? The number of species described since Linnaeus' time has increased so rapidly that those who described them find that they are being accused of creating imaginary species and, in some cases, of endlessly proliferating names. At the same time, others do the opposite when they use the same name to refer to forms that are known to be very diverse and are not linked by intermediate forms. As a result, the species has become a vaguely defined group of more or less similar individuals. One cannot fail to be struck by all the arbitrary delimitations placed on these groups, but any attempts to fix the limits more firmly have run up against so many difficulties that everyone was defining species in different terms. It was necessary to find a common ground that was not based on external characteristics, such as those that Klein said we should use exclusively, and not on anatomical characteristics of the kind Linnaeus was beginning to adopt. Rather, it should be based on purely physiological characteristics that could often be determined only by impractical tests. Aristotle used a simple criterion based more on common sense than on personal observations: he judged the identity of uncertain forms according to the sterility of their unions versus their ability to produce progeny with the same characteristics.

Linnaeus' predecessors had placed species that shared certain similarities into more or less extended groups that, if they gave them any name at all, they designated as genera. Linnaeus was the first to define different degrees of resemblance: in his works, the most closely related species were grouped in *genera*; genera that shared a number of characteristics were placed in *orders*, and the orders were divided into *classes*. The mutual relationships of these various divisions were illustrated in the following table which places terms with a similar hierarchical level in the same vertical column:

Class	Order	Genus	Species	Variety
Broadest genera	Average genus	Most limited genera	Species	Individual
State	County	Town	Neighbor-hood	House
Regiment	Battalion	Company	Squad	Soldier

The last edition of the *Systema naturae* was published in 1766. In 1780, Batsch introduced another term, *Family*, between the order and genus, and the usefulness of the new term is now generally accepted. This orderly gradation of the degrees of resemblances between different groups of animals is a graphic way of illustrating differing degrees of genetic relationships. Linnaeus had already used the nomenclature of political divisions to designate members of the same group in a way that implies that they share a common lineage. The word *family* chosen by Batsch suggests that he had the same sort of comparison in mind, and the word *tribe* that has been employed more recently indicates a similar connection. These may not have been conscious analogies; they are suggested by the very nature of the phenomena we are trying to explain. We note the various degrees of similarity between animals, just as we see a decreasing similarity between members of the same family of humans with increasing distance from their common roots. We compare these two relationships, but instead of representing the classification of animals as a genealogical tree with multiple branches, we view the relationships, either as Linnaeus did, as analogous to those between communities, towns, and provinces on a geographical map or, as Bonnet did, as links in a chain or the steps of a ladder. This concept of an ascending scale of living creatures, which was first proposed by Leibnitz, has had a marked effect on later thinking. Now that it has endured so many years, we should examine how it was presented by the person who was its most ardent champion, Charles Bonnet.

Chapter VI

Philosophers of the XVIIIth Century

Charles Bonnet: the scale of living creatures; global revolutions; the past and future states of plants, animals, and humans; the preformation [emboîtement] of germ cells – Robinet: his ideas about evolution. – De Maillet: fossils. – Erasmus Darwin: transformation based on epigenesis. – Transformation of animals under the influence of habitat; analogies between Lamarck and Darwin. Maupertuis: the roll of matter in transformations. – Diderot: the life of species and the life of individuals.

Linnaeus was primarily a scientist. Although he made brilliant excursions into philosophy, he made a point of restricting himself to the study and contemplation of nature. Charles Bonnet, on the other hand, was primarily a philosopher who questioned nature in order to find problems that called for solutions; he conducted experiments and observations that led to discoveries that had implications at the highest levels of metaphysics. As a philosopher, Bonnet was a fervent disciple of Leibnitz: all his efforts were aimed at demonstrating the possibility of applying to physical matter, including even immaterial things, Leibnitz's *law of continuity* that, as we have already seen, had been adopted by Linnaeus. For Bonnet, all worldly things form an unbroken continuum beyond which there is nothing but God. Minerals grade into organisms, and the latter are linked to one another by a multitude of imperceptible transitions. The divisions we set up in studying the natural world are artificial; in reality, there are no sharp boundaries. Species are strictly linked to one another through the innumerable variations that their individuals can possess: 'A higher intelligence than ours may reveal greater differences between two individuals in one of our categories than we are able to see between two members of distantly related genera. There is about as much variation among creatures of the higher orders as there is among individuals of a simple species. Within every realm of nature, variations are found at every level starting with the atom and culminating with the highest of the angels.'[1] Bonnet believed that there are several habitable worlds and that these worlds have reached different levels of perfection some of which are inferior to ours and others superior.

'Terrestrial beings range themselves naturally into four general classes: *1. the most elementary matter with little or no organization. 2. organized, inanimate objects, 3. organized, animated forms of life, and 4. organized, animated life that*

[1] Ch Bonnet, *Contemplations de la nature*, 1764, Amsterdam, vol. I, p. 29.

A. McBirney, S. Cook and G. Retallack, *The Philosophy of Zoology Before Darwin: A translated and annotated version of the French text by Edmond Perrier*,

is capable of reason.[2] The other worlds do not have the same varieties of life that we see in ours; each has its own particular conditions, natural laws, and organisms. The first and second classes are all that are found in worlds that are much less perfect than our own. On the other hand, there may be worlds that have reached such perfection that they have only creatures of the highest order. Worlds of that kind would include highly organized crystals, sensitive plants, animals that can reason, and humans that are angels.[3]

'How wonderful it is in Celestial Jerusalem where the angels are the least of the intelligent Beings!'[4]

We see Bonnet passing from science to theology and from material beings to the spiritual. These attempts to use his inspirational thoughts to construct a law of continuity among celestial creatures may seem rather naïve to us today. His application of principles based on studies of the tangible world to another world that is completely beyond our perception led him to conclude that there is no distinction between our dreams and imagination. Nevertheless, his use of this same principle to identify the mutual relations of organized life had interesting consequences. After making a careful comparison of plants and animals, Bonnet arrived at the same conclusion that Claude Bernard laid out so eloquently in the last years of his life when he stated that there are no absolute distinctions between the two great realms of organized life: 'Tell an uneducated person that philosophers can scarcely distinguish a cat from a rose bush, and he will laugh and ask whether anything in this world could be more easily distinguished.' Ordinary people who are not accustomed to abstract thinking consider such ideas bizarre and say that philosophers have an unrealistic view of the world. When they ignore the essential properties that constitute the cat and rose bush, they are left with no characteristics that would enable them to distinguish one from another...[5] Plants and animals become nothing more than modified versions of organized matter. They all share a single essential feature, and their distinctive attributes are unknown to us.[6]

By this reasoning, a plant is simply an inferior kind of animal, and one passes by degrees from the human to the animal, from animal to plant, and from plant to mineral. Many of these gradational features still remain to be discovered; Bonnet summarized those that he believed fit this plan in his famous scale [échelle] or 'ladder of life', which I reproduced in full below:

---

[2] Ch Bonnet, *Ibid.*, p. 21.

[3] Bonnet's theory resembles that of Theilhard De Chadin (1959), a Jesuit geologist and anthropologist who believed that man has not only evolved through time but will continue to do so until he converges with divinity. (Trans. note)

[4] Ch Bonnet, *Ibid.*, p. 25.

[5] Ch Bonnet, *Contemplations de la nature*, vol. II, p. 74.

[6] Ch Bonnet, *Ibid.*, p. 77.

HUMANS
Orangutan
Monkey
QUADRAPEDS
Flying squirrels
Bats
Ostriches
BIRDS
Aquatic birds
Amphibious birds
Flying fish
FISH
Crawling fish
Eels
Sea serpents
SERPENTS
Slugs
Cockles
SHELLFISH
Tube worms
Ringworms
INSECTS
Gall insects
Tapeworms and nematodes
Sea Urchins
Sensitive insects

PLANTS
Lichen
Mosses
Mushrooms, agaric
Truffles
Coral and coralloids
Lithophytes [lichen]
STONES
Magnets
Talcs, gypsum,
Selenites
Slate
Figured stones
Growths of crystals
SALTS
Vitriol
METALS
Semi-metals
SULFUR
Bitumen
SOILS
Pure soil
WATER
AIR
FIRE
More subtle material

He linked the creatures in this long list on the basis of purely superficial resemblances. It is difficult to believe that this is the work of a person like Bonnet, a wise and perceptive observer, who was in some ways the equal of Réaumur and Trembley, a careful experimentalist who studied the parthenogenesis of the plant louse and the reproduction of the naidid worms. This is the same Bonnet who carried out such admirable studies as the restoration of mutilated earthworms and the reproductive phenomena of fresh-water bryozoans and ciliophoran protists (Vorticell and Stentor). He did this outstanding work even though he does not seem to have appreciated that the relationships between living creatures are based primarily on their anatomy. He was not concerned with the details of classification but simply looked at the animal kingdom as a whole without considering how secondary groups are related to one another. He discussed at great length a question that Linnaeus considered resolved *a priori*: have the creatures that populate the Earth always remained essentially the same as those we see today?[7] With a remarkable

[7] Ch. Bonnet 1768 *Palingénésie philosophique, ou idées sur l'état passé et sur l'état future des êtres vivants*.

independence of mind, Bonnet ignored all the restraints that the book of Genesis had imposed on Linnaeus. In his view, the world has been the scene of innumerable revolutions, and we have not necessarily seen the last of them. The chaos described by Moses is the result of the most recent of these. But the act of creation, he told us, was another matter. As Whiston had already said, it was a resurrection of animals that had previously been destroyed. Just as the world that preceded Genesis was very different from that of today, the ancient animals had little in common with the ones we know today. Those that will inhabit our planet after the next revolution predicted in the Bible will also be different from those of the two preceding periods. In each global revolution the living creatures underwent profound transformations, and at the end of each period, all life was destroyed only to be replaced by new forms. Strictly speaking, there are no new creatures: the new animals arise from the germ cells contained in their ancestors, and it is these supposedly indestructible germ cells that establish a link between the fauna and flora of each period and the one that follows. What are these germs? How do they modify the living forms? This is what we shall consider next.

We should recognize at the outset that Bonnet's idea of transformism had little in common with the modern version of evolution. He said in Chapter IV of the *Philosophy of Palingenesis* that when 'the embryo of a chicken first becomes visible in an egg it appears in the form of a very small worm,' and 'with improved knowledge and better instruments we shall be able to go back farther into the origin of the chicken and no doubt find it in even greater disguises.' He also noted that 'the different forms at successive stages of development can reveal the series of changes that organized bodies have had to go through in order to arrive at the form that is familiar to us today,' and finally that 'all this helps us foresee the new forms that animals will take on in the future.' If one can judge from these statements, Bonnet already imagined a kind of parallelism between the embryonic development of the individual and the progressive changes of the species to which it belongs, but his philosophical concepts of the development of living forms offered no insight into the origins of organized life. There is complete harmony between the various parts of a single animal, and these parts 'are obviously conspiring to attain a common goal: the formation of this creature that we call an animal, an organized assemblage that lives, grows, feels, moves, and propagates itself.' One can only conclude, Bonnet wrote, 'that such a highly complex and yet harmonious creature could not have been assembled, like a clock, from interrelated pieces or from the interaction of an infinite number of differing molecules joined in successive linkages. A creature of this kind bears the indelible mark of a product created with a single stroke.'[8] Thus, Bonnet pronounced himself opposed to all attempts at mechanistic explanations of the animals; he showed himself to be a resolute adversary of epigenesis and maintained that all living creatures came from a germ cell that was already organized.

---

[8] Ch. Bonnet 1783 *Palingénésie philosophique. Œuvres complètes*, vol. VII, p. 65 ed. of Neufchâtel.

This kind of reasoning has been used by those who attempt to demonstrate the impossibility of evolution on the basis of what is often a very effective adaptation of animals and plants to their individual habitats. When these questions are examined in a superficial way with preconceived ideas and a determination to ignore the fundamental attributes of animals and plants and the admirable harmony in which they co-exist in nature, one can reach only one conclusion: they have been planned and organized down to the smallest details by an intelligent being of infinite wisdom and unimaginable foresight.

The hypothesis of pre-existing germ cells led Bonnet to the reasonable conclusion that there could be no unplanned generation of new forms of life. He was astonished that Rédi had been able to accept such an origin for the worms he found in fruit and in the intestines of animals when there were many more natural explanations for their presence in these places. Moreover, numerous observed facts seem to speak 'in favor of the transmigration of the tapeworm.'[9] Intestinal worms, like all living beings, come from a germ that Bonnet thought of as having 'all the pre-ordained and pre-formed parts essential to the existence of a plant or animal.' The egg, despite the extreme simplicity of its composition as we understand it today, fits this explanation perfectly,[10] particularly when Bonnet added that one must not imagine that 'all the parts of an organized body are present in miniature in the germ cell exactly as they will appear in full scale in the developed whole.'[11] But in saying this he makes a concession to the numerous, well-known examples of the metamorphism of insects. In essence, Bonnet saw in the germ cell a very complex organized being, and he was obviously pleased whenever he could show that an egg or an embryo contains elements that one would not have expected to find in them.

Bonnet believed that these germ cells, which are almost as complex as the adult animals, could only be formed, as animals are, in a single stroke in the act of creation. He agreed that they have been created, as Vallisneri had first supposed, as a complete assemblage that was incorporated into the structure of a living body where they awaited their turn to develop and grow.

In other words, there is no such thing as *reproduction* in the sense of producing a new form of life from a progenitor, only *development* of a pre-existing germ. To postulate that the germs of living creatures are often enclosed one within another implies that the most elementary units must be of such disproportionately small size that they are hard to imagine. But that was no problem for his kind of reasoning. Before any objections of this kind could be raised, Bonnet put them off by saying that the doctrine of preformation [emboitement] appeared to him to be 'a beautiful triumph of pure reason over the senses. I have shown how absurd it is to oppose this hypothesis with calculations that do nothing but play on the imagination.

---

[9] Bonnet, *Considerations sur les corps organisés, Œuvres complètes*, vol. III, p. 37 and 38.

[10] Bonnet, *Œuvres* vol. VII, p. 68.

[11] *Ibid.*, p. 67.

Clear reasoning easily reduces such objections to their true value... Imagination can mislead us with all sorts of false images and should not be allowed to influence our judgment of things that can only be properly perceived by a philosophical eye.'[12]

This distinction between empirical and philosophical reasoning assumes that the senses can be deceptive, but that one can never be misled by pure reason. Once this kind of reasoning is accepted, facts are no longer an embarrassment. The most impressive example of epigenesis was seen in vegetation where branches, stems, and leaves were said to be independent individuals that even our empirical eye can see growing one upon the other. Bonnet's concept of plants differs quite materially from our own. 'A tree,' he says, 'is not a single, unique entity. It is composed of as many component trees and bushes as it has stems and branches. All these lesser trees and bushes can be thought of as being grafted upon one another and linked to the main tree by an infinite number of connections. Each secondary tree, each shrub, each bush, has its own organs and life. It is a complete, small individual that represents in miniature the grand total of which it is a part.'[13] Other organisms resemble the plants in this regard: the polyps, the budding of which has been studied by Trembley, the tapeworm, composed of repeated segments similar to its own, the naidid worms, the tube worms, the earthworms whose modes of reproduction and segmentation Bonnet studied so thoroughly; all these have a similar development. They are true 'zoophytes.' The same explanation can relate the reproductive phenomena of the zoophytes and plants to the theory of preformation: the germs are distributed throughout their entire bodies and transform it into a kind of 'universal ovary.' In a growing plant or in a budding polyp, these germs develop spontaneously as individuals, and they can remain united or become separated. In worms, it takes an accident of some kind to cause them to develop because in this case the parts can become new individuals only after being separated from one another. Thus, thanks to the hypothesis of invisible germs [preformation], the most conspicuous facts of epigenesis are used to advance the theory of evolution.

One can endow these invisible bodies with any kind of properties one likes without fear of being discredited by anything tangible we can observe. Bonnet believed that all his invisible germs were equally indestructible. When a living body, even an egg, is destroyed, the indestructible germs that it contained are liberated and lodge themselves wherever they can find a suitable place to reside. 'Indestructible germs can be dispersed quite easily into any of the various bodies that surround them. They can stray into one body or another and remain there until it begins to break down, then pass unchanged from that body into a third, and so on. I have no trouble imagining that the germs of an elephant could lodge themselves in a particle of soil, then pass from there into a ripening fruit, from there

---

[12] *Ibid.*, p. 152.

[13] Ch. Bonnet, *Palingénésie philosophique; Œuvres*, vol. VII, p. 163.

into the body of a fruit fly, and pass in this way from one thing to another.'[14] These germs, which were created at the time our world came into being, 'survive for centuries in all sorts of changing conditions.' Nothing can overcome the strength and durability that 'enables the first germs of organized beings to survive until the last day of our planet.' Just as Leibnitz envisaged an inherent harmony between our mind and the movements of our body, Bonnet believed that our actions always correspond to our innermost thoughts, and in the same way he perceived a perfect parallelism between the astronomical system and organic life or between the conditions that make our planet habitable and the creatures that populate its surface. The germs created for each period of time remain hidden in the organisms that shelter them and wait there until conditions permit them to emerge again. In this way, the beings of each geological period are linked to those of an earlier one that harbored their germs, and they are independent of the prevailing conditions because all the germs were created at the same time. Thanks to the harmony established between the evolution of organic germs and that of our planet, new fauna and flora appear without any need for a new act of creation.

Though not one to shirk difficult tasks, Bonnet decided that he should confine his considerations to certain limited periods in the Earth's history: that which preceded the revolution described in Genesis and that which will follow after the world's end in the conflagration forecast by the prophets. He had a rather strange view of the forms that animal life would take in the future. The germs from which they will arise could survive the fires only if they were composed of a special inflammable material, some special kind of ether. 'If we start with the assumption that a small ethereal body encloses in miniature all the organs of the future animal, it would seem that the bodies of newly created animals would be composed of a rare type of material organized in a way that they will be protected from the changed conditions that affect larger bodies and make them so vulnerable to destruction. Being largely immune to these dangers, the new body will not have to heal as many injuries as the original one does. Its built-in mechanisms will be much superior to those we admire in existing species, and there is nothing to indicate that in the future animals will need to develop in the same way they do now.'

With this, we come to the realms of the mind and immortality, and we find ourselves dealing with total fantasy. Here we have one of the most ingenious minds of his time, a keen observer who was capable of rigorous reasoning, using a handful of obscure observations, boundless imagination, and a literal interpretation of the Bible to construct fantasies for which there was no objective evidence. Instead, he discounted any evidence that was not in accord with the deductions he thought were based on his reasoned judgment.

Bonnet was not the only philosopher who followed such a path. At that time, many scientists, philosophers, as well as pure dreamers, were preoccupied with the origins of animals and man.

---

[14] Ch. Bonnet, *Palingénésie, Œuvres*, vol. III, p. 152.

Robinet, in his books *De la nature* (1766) and *Considérations philosophiques sur la gradation naturelle des formes de l'etre* (1768), put forward ideas that, although they were ridiculed by Cuvier, were not very different from Bonnet's. Like Bonnet, Robinet adopted an extreme interpretation of Leibnitz's law of continuity and proposed that there is life in all matter, that the stars, Sun, Earth and planets are all living creatures, that all beings make up a continuum, that there are no classes, orders, genera, or species but only individuals that are identified only by imperfections that we mistakenly take as evidence for specific identities. Individuals are born from germs that undergo progressive development; nature forms them directly. The material world is governed by invisible forces. Nature never repeats itself, and the time may come when there will not be a single being with the form it has today. Living forms develop by a process of perfection from the simple to the complex. There could be immaterial creatures superior to humans, but man is part of an infinite chain consisting of gradational variations of a simple prototype. All intermediate forms are works of nature leading up to humans, its most perfect creation until now. This masterpiece of nature will be further perfected in the future if man, once he becomes hermaphroditic, reunites the beauties of Venus with those of Apollo. In the end, this perfection of humanity is not much stranger than that which Bonnet dreamed of.

De Maillet, better known under his chosen pseudonym, Telliamed, followed Bonnet and Robinet in looking for the explanation for the origin of life in the creation of infinite numbers of germs, but he made the sea the reservoir in which they reside. All animals, including even humans, were once primitive marine animals. De Maillet believed that the sea had a much vaster extent in the past and cited as proof of this the enormous numbers of seashells that are now found on land, even on the highest mountains. As the continents grew larger, some of the marine animals were accidentally stranded on shores where there was still a certain amount of moisture, and from there they adapted themselves to dry land. The individuals stranded in this way became accustomed to the new kind of life that these conditions imposed on them and transmitted to their descendants the newly acquired features of their organs. There is no need to dwell on the bizarre arguments that de Maillet employed to support his hypothesis, but we must give him credit for having recognized the true nature of fossils and for having seen their importance at a time when most scholars still refused to accept them as remains of creatures that had lived in the past. He also deserves credit for having thought that living organisms are susceptible to modifications that they can transmit to their descendants. This led him to recognize the importance of the well-known but neglected phenomenon of heredity.

By accepting the possibility of hereditary changes in the structures of living beings, de Maillet advanced a step beyond Bonnet and Robinet who saw such modifications as nothing more than a continuation of the original miracle of creation. Dr. Erasmus Darwin, grandfather of the illustrious advocate of transformism, went a step farther than de Maillet. In his *Zoonomia,* he laid out a system in which he used very insightful arguments to support ideas that differ little from those that would later be developed by Lamarck. In an effort to make his system intelligible, Erasmus Darwin started by trying to understand the development of the embryo and postulated that, with time, the species that came from it had undergone an analogous evolution of much longer duration. He rejected the doctrine of an embodiment of germs that assumed the existence of living bodies infinitely smaller 'than the devils that tempted Saint Anthony and of such size that 20,000 could comfortably do a song and dance act on the tip of the finest needle.' In his view, the embryo was a filament that was probably the extremity of a motivating nerve fiber. This filament possesses certain properties: some are distinctively its own while others have been transmitted to it from its parents. In the latter case, it is really only a branch or prolongation of its parents, because it began as part of their substance. The embryonic filament is sensitive to stimuli and has a will of its own. It also possesses the ability to nourish itself, and it grows larger, more complex, and more perfect by adding new parts from other living matter that it makes part of itself. At first this living matter is added through the influence of the inherent properties of the embryonic filament, but its new organs bring new faculties that create additional needs of their own. These needs determine a way of life and behavior that contribute to the further development that each individual undergoes in the course of its existence.

The progress of evolution of species was thought to have advanced in a similar way: living organisms were first created as extremely simple forms reminiscent of those seen in the earliest embryonic stage of development of the individual. They started with a very small number of species, and, just as each chemical molecule has inherent properties that determine the sorts of compounds it can produce under different conditions, the living primitive filaments were similarly endowed with distinctive faculties that determined in large measure the course of their ultimate development. Given the resemblances that all warm blooded animals share, it is probable that all these animals have descended from the same kind of primitive ancestor, possibly the same ones have given rise to other red- but cold-blooded animals. The distinctive characteristics of fish suggested that they should be assigned a special origin all their own, but the intermediate forms that link them to warm-blooded animals indicate that they are more closely related to the latter.

'Insects without wings, such as the spider, scorpion, or crab louse differ in many ways from those with wings like the mosquito, termite, yellow jacket, or dragon fly, and are so distant from red-blooded animals in both their bodily forms and ways of living that it is hard to think that they could all stem from the same progenitor as that which produced the various classes of red-blooded animals … Another class of animals that Linnaeus designated as worms has a more simple structure than those mentioned earlier. This simplicity, however, poses no argument against the hypothesis that they have been produced by a single living filament.' Erasmus Darwin considered the vertebrates, articulates, and worms as three organic types that developed simultaneously and in parallel from organic forms that were equally simple but endowed with different properties.

Even though the three lines recognized by the English scientist do not correspond to what we now know about their genetic relationships, the basic idea that several organic types were generated and developed independently must still be the only form of transformism that could be in accord with the facts of paleontology. The reduction of all these animal forms to three distinct lines shows that, as early as 1794, several years before the publication of the first works of Cuvier, Erasmus Darwin had already recognized the close relationships of animals making up Linnaeus' first four classes and the important differences that separated them from those of the fifth class, but the English philosopher left the sixth class in a chaotic state that Cuvier managed to straighten out a few years later.

Each of the living protozoa that were to become the roots of the three major animal lineages was destined to develop according to the particular properties with which it had been endowed, but in each particular case its evolution was regulated in part by the animal's reaction to favorable or unfavorable sensations and by its efforts to augment its well-being or reduce its discomforts. The copious amounts of water and air that were always available served as a means to meet three basic needs essential to development: the need to reproduce, the need for nourishment, and the need for security. The evolving creatures acquired the necessary means to defend their habitat, sources of food, and breeding partners. When Erasmus Darwin described this process of evolution, he came close to enunciating the concept of the struggle for life and natural selection. Speaking of conflict between males competing for the same female, he said: 'The final cause of this contest amongst the males seems to be that the strongest and most active animal should propagate the species, which should thence become improved.'[15] It must be emphasized that, instead of saying the *cause*, Charles Darwin would have said the *consequence,* but the grandfather and grandson were clearly in agreement on the reality of natural selection: Erasmus Darwin, like Lamarck, believed that animals acquire organs in order to satisfy some vital need. For Charles Darwin, these organs appeared accidentally, and natural selection preserves and perfects those that prove useful while allowing those that do not to atrophy. In this way, plants and animals adapt themselves to prevailing conditions without these conditions acting on them directly. The individual simply submits to the necessity of adapting to its environment.

As ingenious as it was, the hypotheses of Erasmus Darwin left us totally ignorant of the basic reason why these organisms appear in the first place. He takes us back to the creation of the primitive living filaments and leaves us there. Many thinkers of the XVIIIth century would have found this solution inadequate. Already, in the XVIIth century, Descartes had tried, without success it is true, to explain the formation of animals and humans by a single process. Maupertuis[16] recognized the difficulty of doing this and suggested that there could be only two possibilities: matter could be endowed with special properties that, once acquired,

[15] *Zoonomia* vol. I, p. 507.

[16] I am indebted to my venerable friend, M. Victor Considérant, for drawing my attention to these passages from the works of Maupertuis.

would render that individual capable of producing offsprings with the same physical and mental faculties, or, alternatively, that all plants and animals have existed since the Earth began, and that what we think are the new features of the genus, are simply the result of growth and development of parts that were too small to be recognizable. This latter idea is the system of preformation [emboitement] of germs adopted by Vallisneri, Leibnitz, and Bonnet.

'This system of simultaneous formation that requires nothing more than the growth and development of parts that already existed in the fully-formed individual was thought to resolve all the difficulties. The only thing one had to know was where to store these inexhaustible resources. Some placed them in one sex, some in the other, but most remained content with the basic ideas.

'If one examines this system more closely, however, one sees that it really explains nothing. To suppose that all the individuals were formed by the Creator on the same day is to fall back on a miracle. It is not a physical explanation. We gain nothing from this simultaneity, because events that appear successive to us are always simultaneous for God.'

Having rejected the doctrine of preformation of the germ cells, Maupertuis aligned himself with the doctrine of transformism, but he had his own particular interpretations. By means of a process that is common among theoreticians and is just a device for putting the mind at rest rather than offering a true explanation, he attributed very important intellectual properties to this invisible material: desire, aversion, memory, habits, and so on. He then called upon these properties to deduce a whole system of evolution.

'The essential elements of the fetus swim in the seeds of its father and mother. Each part preserves a kind of *memory* of its earlier place in the parents and, when able to do so, tends to take on the same form in the fetus. This orderly progression is responsible for the conservation of the species and the resemblance of the individual to its parents.'

In certain ways, this is the hypothesis that Charles Darwin referred to as *pangenesis* when he revived it in his book on the *Variations of animals and plants under the action of domestication.*

Maupertuis added that, 'if some of these genetic elements are missing from the seeds or fail to assemble in the proper way, the result is a monstrosity in which some parts of the body are missing or malformed. If the bodily elements are disproportionately large or remain undeveloped so that another can take its place, the malformed creature is born with superfluous parts.

'If the elements come from different species of animals that still retain a close relationship in which some elements are more strongly attached to the father and others to the mother, the result of their union will be a hybrid…

'It is rather common to see an infant resemble one of its ancestors more than its immediate parents. The elements that form some of its features may have preserved the essentials of their forbearer's condition rather than that of the immediate parent, either because they may have resided longer in one than in the other or had a dominant ability to unite. They will then be manifested in the fetus much as they were in the ancestor.'

There are other explanations for heredity, atavism, or hybrid characteristics that differ little from those that have been provisionally set out by the recent work of Charles Darwin. But Maupertuis expected his hypothesis to yield the explanation for the origin of new species.

'Shouldn't one be able to explain,' he asked, 'how two individuals can produce so many dissimilar descendents? They must owe their origin to some fortuitous circumstances in which their elementary parts no longer play the role they did in the parents. Each deviation could have made a new species, and repeated divergences would bring about the infinite diversity of animals that we see today, a diversity that may still be increasing with time but at a rate that may be scarcely perceptible over the course of the centuries.'

This is a clear statement of the theory of heredity. Maupertuis has even tried to attribute the strange sterility of hybrids to a kind of incompatibility that comes from the innate differences of species and prevents them from varying beyond certain limits. He never really tells us how these differences have been acquired, and it was Charles Darwin's innovative work that demonstrated for the first time that it was a consequence of the principle of natural selection that had scarcely been perceived until then.

Maupertuis also believed that the manner in which plants and animals develop is not significantly different from that which we see in the growth of crystals. Thus, the living world and the world of minerals share certain traits, which, if one includes such faculties as sensations, memory, likes and dislikes, are normally considered characteristics of the most advanced living creatures.

In his *Pensé es sur Lînterpretation de la Nature,* Diderot[17] discussed a dissertation that Maupertuis published in 1751 under the name of Doctor Baumann d'Erlang. Maupertuis did not address the question of how we know whether material is inert or living or whether inert material can spontaneously come to life, but he thought that one could explain animals by simply endowing organic molecules with a kind of rudimentary instinct that drives them ceaselessly in search of the conditions into which they can fit most comfortably. Thus, an animal is 'a system of different organic molecules that joined with one another in response to an impulse similar to the crude sensations with which it was endowed at the time it was given its basic material substance. It continued to react in this way until each part found the place that was best suited to its particular shape and function.'[18] This place, which is favored as being the most suitable, can be altered by the innumerable disturbances that are always being imposed by other organisms that have not yet reached a state of repose in relation to the molecules. It is also true that Diderot wondered 'whether plants and animals have always been and always will be what they are now.' He added: 'Just as a plant or

[17] The State Councilor of Mesnil and M. Victor Considérant have both pointed out to me the transformist opinion that Diderot expressed on several occasions.

[18] Diderot, *Pensées sur l'interprétation de la nature*. vol. LI, 1754.

animal first begins to grow and gain strength, then declines and finally passes away, could not an entire species do the same? If faith tells us only that animals have come from the hands of the creator more or less as we see them and that we are not permitted to have the slightest doubts about their origins and ultimate end, the philosopher is left with nothing but his speculations. Might he not suspect that throughout time the distinctive elements of animal life have been dispersed in a confused mass of matter and that these elements came to be assembled simply because it was possible for that to happen. They progressed by infinitely small steps from the embryonic state to one of increasing organization and development and gradually acquired mobility, sensations, ideas, thought, reflections, conscience, sentiments, passions, gestures, articulated sounds, language, laws, arts and sciences. They continued to evolve over periods of millions of years and may continue to develop in the future in ways that we cannot possibly imagine. We have no way of knowing whether it has always had or will continue to have a static state or whether it may escape from this state by a process of deterioration during which its capacities will leave it in the same way that they came in. Might it not disappear from nature or continue to exist with a form and capacities that are entirely different from the ones we see in it during this brief instant of time?'

In addition to naturalists and observers like Linnaeus, some of the philosophers of the XVIIIth century were beginning to address the explicit problem of gradual transformations of the species. None of them managed to resolve it, but another naturalist, as highly endowed as Linnaeus though in a different way, and equally free of dogmatic prejudices and preconceived ideas, embarked on a course that would later be followed by an unbroken succession of brilliant disciples. This man was Buffon. With him, zoological philosophy embarked on a new era. Henceforth, there would be a new precision, and greater progress would be made toward truth in a half-century than in all the centuries that had passed since Aristotle.

Chapter VII

Buffon

Buffon's opposition to classifications because they necessarily lead to transformation. – The utility of artificial systems. – Geographical distribution of animals. – Probability of modifications in species. – Extinct species: the struggle for life. – Opposition to the doctrine of ultimate causes. – The principle of continuity.

Buffon's work was inspired by a conception of life that was quite different from that which Linnaeus' work had developed so well. Linnaeus' classification was, in a sense, a way of summarizing all zoological thinking. He devoted all his efforts to research based on the *natural method* and made this his supreme goal. He saw nature as immutable, and concluded that there was therefore nothing to explain. The naturalist should simply try to understand the design of creation and attempt to reproduce the plan within the system on which it was based. Buffon put aside the orderly divisions and subdivisions in which the students of Linnaeus tended to view science. He studied each animal species independently, and, unlike his illustrious Swedish predecessor, who ignored the true nature of species in order to use dogmatic definitions, he preferred to leave the door wide open to further studies and new interpretations. He began by asking himself whether or not species can vary and, if they can, why they do so, and what limits are placed on those variations.

Various explanations have been offered for Buffon's aversion to classification systems. The president of the council, Lamoignon de Malesherbes accused him of rejecting them because he simply did not understand them. Daubenton also argued that he had not properly understood Linnaeus' method. Flourens agreed with all these reproaches and suspected that Buffon was bitterly jealous of the great Swedish naturalist. Despite the authority attached to these three names, two of whom were among Buffon's most eminent friends and collaborators, one would hope that there is no compelling reason why we should accept their view. To reproach a man with Buffon's knowledge and great intellect for rejecting a system because he does not understand it seems illogical when one considers that when Linnaeus first presented his *system of nature* it was far less complicated than it has become today. It might take a person like Buffon a while to fully comprehend the system dealing with mammals, but it is hard to imagine that he would not explore the topic when he was starting on his *Natural History* and would not see that it was necessary to understand its bearing on his own work. On the other hand, Buffon was endlessly correcting himself, always trying to make his ideas clearer and more

A. McBirney, S. Cook and G. Retallack, *The Philosophy of Zoology Before Darwin: A translated and annotated version of the French text by Edmond Perrier*,

precise, and abandoning ideas that no longer appeared correct in order to resurrect others he had rejected earlier. There was no false modesty in his efforts to lay out for his readers the innermost thoughts going through his mind. Would he have allowed pride to make him condemn methods that he recognized as genuinely scientific? As for the accusation of jealousy, the Count de Buffon had wealth, honors, and glory and was universally considered a scholar of the first order, a writer of genius living in a beautiful capital city where he was admitted to the most brilliant court of Europe. How could he envy a professor of the University of Uppsala who, though no doubt famous, had a much more modest status than the noble academician, manager of the king's botanical garden and the museum of natural history in Paris? Can we believe, as Daubenton did, that Buffon had not understood Linnaeus' method when he wrote: 'To class man with the monkey, the lion with the cat, to say that the lion is a cat with a mane and long tail is to degrade and distort nature instead of describing it and naming it?'

'Buffon,' said Daubenton, 'wanted to cast ridicule on the naturalists for putting the cat and the lion in the same genus. Of course, a cat is not a lion, and that is not what Linnaeus meant to say. Buffon did not fully understand Linnaeus' method; if he had just gone through the list of species under the genus known as *felis*, he would have found that there is a lion species and a cat species… This error resulted from the manner in which genera are named for a species that is one of its best-known members.' Future developments showed that Daubenton underestimated Buffon's appreciation of the necessary consequences of Linnaeus' system and of classifications in general; Buffon may even have understood better than Linnaeus himself the direction that nomenclature could lead studies of zoology. It was these consequences that Buffon dreaded, at least at that time. He stated this clearly in terms that showed that his opposition to Linnaeus was based on far higher motives than those expressed by Lamoignon de Malesherbes and Flourens.

Before undertaking his work on animals, Buffon wrote about the natural history of man with a breadth of view that, even to this day, remains unequalled. He placed man so high in nature that he made him something approaching a deity. One of the immediate consequences of the classification systems was to put man back into the animal kingdom. For Linnaeus, man was only the highest member of the order of primates and was still close to the monkey. On the other hand, in their desire to express the various degrees of resemblance of animals, the students of Linnaeus had compared living beings to a large family and established a more tangible sense of the similarities of animals of the same group. This allowed us to speak of the different divisions of the animal kingdom in the same terms we use to designate an assemblage of humans with different levels of relationship, such as those expressed by words like *family* and *tribe*. If one takes the word *genus* literally, it could be applied only to animals having a common progenitor.

Linnaeus and his disciples used simple comparisons and metaphors to render their precise method of arranging animals more intelligible. He 'counted as many species as couples came from the hands of the creator' and accepted the immutability of nature as axiomatic and saw no danger in this. Influenced much less by the Bible than by his studies of the Earth and man, Linnaeus was already thinking in terms of gradual modifications of the world and its species. Buffon, however, insisted that things have not happened as simply as Linnaeus liked to portray them. He feared that people with very adventurous minds, might give in to a line of reasoning that he had already begun to experience himself and, instead of scrutinizing the origins of living creatures, would take Linnaeus' portrayal of them too literally. They assumed that any animals that the nomenclature system placed in the same family must be united by their blood lines. By this reasoning, man would become a cousin of monkeys, and Buffon recoiled before the enormity of this conclusion. He expressed this view very clearly, and it is rather surprising that the various interpretations offered for his opposition to the Linnaean classifications have been accepted without considering his own explanation, which is the only one consistent with his genius. It is worth citing the passage in which he expressed his way of thinking in this regard. It is found near the beginning of the natural history of quadrupeds and is an introduction to the remarkable chapter he devoted to one of the most humble of our domesticated animals, the donkey.

Buffon says that 'if one considers this animal closely in all its details, it appears to be nothing more than a degenerate horse… One could attribute the minor differences between the two animals to the influence of climatic conditions in the distant past, to their food, or to several fortuitous generations of small, half degenerate wild horses that little by little continued to decline until they became what we take to be a new, stable species. Or there could have been a succession of similar individuals, all corrupted in the same way, and different enough from the horse to be regarded as forming another species. The fact that the color of the horse's coat varies much more than the donkey's indicates that horses have been domesticated for a longer time, because all the domestic animals vary in color more than wild animals of the same species… On the other hand, if we consider the differences of temperament and behavior of the two animals and their inability to interbreed and produce an intermediate offspring that can continue to reproduce, it seems more reasonable that these are two distinct species that have long had the same basic differences they have today… Did the donkey and the horse originally come from the same root? Are they, as the taxonomists [nomenclateurs] say, members of the same *family*? Or have they always been different?

'A physical scientist will recognize the basic nature of this problem and will appreciate its consequences and the difficulties they present. The question of how closely-related animals originate is now being addressed from an entirely new point of view. No matter what animal we choose from the immense varieties that populate our universe to serve as a model – even humans – we find that, despite the distinct identities that they have gradually acquired, *all these creatures share an overall design that we can trace back through a long course of development.* Their basic physical differences evolve much more slowly than their outward appearance, for they include such things as the digestive organs, circulation system, and mode of reproduction, all of which are essential to the animal's existence.

These are the parts of the body that have the greatest influence on external forms, and their striking resemblance from one animal to another reminds us of the basic design on which all of them seem to have been conceived… Consider how essential some of these parts are to the form of individual animals. Take the limbs, for example. We find them in all the quadrupeds, birds, and even fish and turtles. As Daubenton pointed out, the foot of a horse, which looks so different from the human hand, is composed of an identical set of bones, and one might reasonably ask whether this hidden resemblance is not more marvelous than the most conspicuous differences. This constant conformity to a common design extends from humans to quadrupeds, from quadrupeds to cetaceans, from cetaceans to birds, from birds to reptiles, from reptiles to fish, etc., all of which possess the same essential parts, such as a heart, intestines, spinal cord, and nervous system. Does this not tell us that *in creating the animals a supreme being wanted to employ a single idea that was followed in countless variations*, and that this was done so that humans could admire the magnificent execution and simplicity of the design.

'From this point of view, not only the donkey and the horse, but also humans. monkeys, quadrupeds, and all other animals could be considered members of a single family, but does it follow that only God could conceive of such a large, diversified family and create it from nothing? There may be other small families that nature has produced over time, and some of these may consist of only two individual species, like the donkey and horse, while others, like that of the weasel, the sable, the ferret, marten, etc., have a greater variety. Similarly with vegetation, families may include ten, twenty, or thirty types of plants. If such families exist, they could only have formed from mixtures, variations, and degradation of ordinary species. *Once one accepts that there could have been such diverse families of plants and animals and that the donkey belongs to the horse family and differs only because it has degenerated, one could just as well say that the monkey belongs to the human family and is a degenerate human, that the human and monkey share a common origin, just as the horse and donkey do.* By similar reasoning, each family, whether it be one of animals or plants, could have had only one root. *All animals have come from a single progenitor that, with the passage of time, has produced all the other races of animals, either by perfecting itself or degenerating.*

'The naturalists who set up families of plants and animals so casually do not seem to appreciate the consequences of what they are doing when they reduce the original products of creation to a small number of individuals… No, it is certain, as shown *by revelation,* that all animals participated in the act of creation and that the first two members of each and every species were formed by the hand of the Creator. One must believe that at that time they differed little from the descendents we see today.'

This passage is important in more ways than one. First, it is a clear and complete statement of the theory of the unified plan for the composition of the animal kingdom that would later be carried to its ultimate consequences by Geoffroy Saint-Hilaire. The fixity of species, that Buffon later rejected, is affirmed without reservations and in almost the same terms as those used by Linnaeus. And finally, what Buffon objected to was not so much the classifications themselves as the tendency of the classifiers to present their systems as a faithful image of nature. He was especially opposed to the so-called *natural* families, and he rejected them for the simple reason that they were supposed to be natural. They were explained as modifications of one of the species from which they are said to be derived. 'There would be no limits to the power of nature, and one would be able to suppose that, with time, it was possible to derive all organized beings from a single parent.'

Buffon never denied that classification systems could be useful, but he emphasized that 'one must bear in mind that Nature changes in subtle, imperceptible ways and the intervals and magnitudes of these changes are far from uniform. As species become more advanced there tend to be fewer of them, and the differences between them increase. Minor species, on the other hand, tend to be very numerous and differ less from one another, so that one can easily confuse them and mistakenly lump them together in the same families. They are a nuisance because we feel obliged to remember the small differences of so many forms. But one must not forget that we have set up these families as a convenient way to deal with them, and if we cannot understand the true relations of all the forms of life, it is our own fault, not nature's. Nature knows nothing about these so-called families; it is concerned only with individuals.'

This is essentially the line of thinking that Buffon followed in his *Histoire Naturelle des Animaux*. If one does not accept the theory that living beings have descended from a unique primitive ancestor and rejects what we now call *evolution* [*transformisme*], the classifications are seen as nothing but artifacts of our imagination. They are useless if one is concerned with factual details, and they can even be dangerous if, as happens too often these days, they are presented as genuinely scientific. Buffon made little use of them when he was dealing with the large mammals, but he relates similar animals, such as the horse and donkey, or the goat and sheep, the various species of pigs, elk, deer, roe deer, the wolf and fox, the American and Eurasian otters, the marten, the skunk, the ferret, the short-tailed opossum, the ermine, and the donkey [grison], and the various species of rodents, etc. The natural series are perfectly constrained, but Buffon breaks them up in a deliberate way according to his own clearly expressed reasoning. The only time that he followed the classification system almost completely was when he was dealing with birds. The great numbers of birds had to be put into some sort of methodical order so their individual attributes would not be misinterpreted. In this case, it was necessary to make use of the methods set up by the taxonomists [nomenclateurs], and Buffon applied this method so skillfully that most of his natural groups have held up remarkably well.

Buffon's determination not to confine himself to any particular system of classification had some fortunate consequences. He first described domestic animals, then wild animals, first of Europe and then the rest of the old continent, and finally those of the new world. In other words, having no good reason to do otherwise, he proceeded by considering regional *faunas*. In this way, his attention was focused on the general characteristics of each of these fauna, including their geographic distribution, and its possible causes. Buffon deserves to be thought of as the founder of biogeography, but these studies forced him to modify profoundly his ideas about the origin of species. His comparisons of the faunas of the two continents led him to conclude that species could vary, a concept that he had earlier opposed. He became a transformist a century before Darwin, during his famous voyage around the world, would conceive the doctrine that is now immortalized in his name. Darwin came to this view after observing diversified but closely related faunas in different regions of the world and seeing with his own eyes how they could succeed one another.

After showing that only a small number of animals were common to both Europe and America, Buffon pointed out that most European animals have American analogs. He was under the impression that the animals of the new world are smaller than the corresponding ones in the old world, and he summarized this view in saying:

'When we consider the general consequences of all this, we find that man is the only living creature with a nature that is strong and flexible enough to survive and multiply almost anywhere and can adapt himself to all the climates of the world. No other animal has attained this privileged role. Far from being able to spread into all possible regimes, most are confined to certain climates and even to particular regions. Everything about man is the work of heaven. Animals, however, are in many ways the products of their local habitat on Earth. Those of one continent are seldom found on the other, and those that are found there are altered, are smaller, and changed almost beyond recognition. Do we need any more evidence to be convinced that their form is not unalterable, that their nature, being much less constant than man's, can vary and even change fundamentally with time? By the same reasoning, can we doubt that the species that are least perfect, most delicate, most ponderous, least active, and least able to defend themselves have already disappeared or are doomed to disappear in time? Their status, the way they now live, their very existence depend on the way man influences the world in which they must live.'

By this time, Buffon's thinking had evolved to a remarkable degree. He now believed that species could vary, that their status depends on the conditions in which they live, and that we should not assign too much importance to man's influence, because it was important to recognize that species that are less suited to their environment can disappear spontaneously. In this, he foresaw the fundamental importance of *natural selection*: 'The prodigious mammoth no longer survives anywhere. This species was certainly the dominant, largest, and strongest of all the quadrupeds. Many smaller, weaker, and more common animals must also have perished without leaving any sign of their former existence! How many other species have been altered or, let us say perfected or degraded, by the vicissitudes of the land and seas, by favorable or unfavorable changes of natural conditions? How many could still be the same as they were in the past after enduring the prolonged influence of a climate that changes in ways that may either favor or hinder their survival!'

It became evident to him that species had not only disappeared, but new ones had appeared in their place. After having denied this so vehemently in the past, Buffon now conceded that it was indeed true. He had the impression that all the animals of America had developed recently: 'It could well be possible, therefore, that even without altering the order of nature, all the animals of the new world may not be exactly the same as the old-world progenitors from which they could have been derived. One could say that, having been separated by the immense seas or inhospitable lands, they would have been living under different conditions and in a climate that was new to them. With time, those same conditions that isolated them would also lead to qualitative changes of their nature. But that must not prevent us from looking at animals living today and recognizing them as different species. Whether these differences have developed over time as the result of changing climate or differing regional conditions or have existed since the time of creation, they are no less real. Nature, I am sure, is in a state of continual flux, but all we can do is address species as we see them today and look back for clues of what life could have been like in the past or what it could become in the future.'[1]

Buffon was still skeptical of classifications, but it was now mainly because of the way taxonomists had misused them. Instead of looking for possible modifications of a particular form, they continued to proliferate endless species in a vain desire to attach their names to useless discoveries. Buffon was definitely on the path toward conversion. After having been such a firm believer in the fixity of species, he was now beginning to suspect that there was something to be said for transformation. This evolution of his thinking went even farther when he saw that there was no intermediate ground; one could not be a semi-transformist. From then on, he could not justify his opposition to a methodical ordering of the animal world. He wrote:[2]

'In comparing all the animals in this way and relating each one to its genus, we find that the two hundred species that we can now identify can be reduced to a rather small number of families or precursors from which all of them could possibly have descended.

'And, to bring some order to this simplification, we can start by separating the animals of the two continents and observing that all the known animals that are common to the two continents, and even those that are confined to the old world, can be reduced to fifteen genera and nine isolated species.'

Eleven of these genera correspond exactly to our groups of single-toed browsers, the browsers with hollow horns and those with solid horns, the pigs, dogs, civets, weasels [mustélidés], rodents, edentates, monkeys, and bats. The other four are more of a problem: Buffon put cows completely apart, united the porcupines and the hedgehogs, and considered amphibians to be close relatives of the otter, beaver and seal. But apart from that, his groups are delimited as sharply as those of other taxonomies. In fact, the classification of mammals that Buffon proposed was

---

[1] *Histoire naturelle des animaux, Animaux communs aux deux continents.*

[2] *Dégénération des animaux.*

basically a genetic one. In his chapter on the donkey, for example, he made the point that species of the same family can be considered descendents from a common ancestor, and he came back to the idea that several species of the New World had descended from those of the Old World.

In treating genealogy in this way, it is helpful to know the degrees of relationship of the species, and in order to determine this Buffon made special use of the animals' ability to interbreed, and he recommended that future naturalists do the same: 'What better way is there to learn how closely animals of different species are related than by observing what comes from their frequent attempts to interbreed? Is the donkey closer to the horse than the zebra? Is the wolf closer to the dog than the fox or the jackal? At what distance from man do we put the great apes whose bodies are so similar to our own? Were all animal species formerly what they are today? Have their numbers increased or decreased? *Have weak species been destroyed by stronger ones* or by the tyranny of man, whose numbers have increased a thousand fold over those of any other animal species? What similarities can we find in this relationship of species and a more common one we see in the different races of the same species? Does not a new race generally get its start in the same way that individuals produced by cross-breeding of species diverge from the pure species?... There are so many questions that one can ask about all this and so few that are being resolved!' How do Buffon's questions differ from those that scientists are discussing so passionately today? Although he was sometimes plagued by serious doubts, Linnaeus was not particularly concerned about the nature of species, whereas Buffon considered this the greatest enigma that nature posed to human intelligence and strove to resolve it. It is to Buffon's credit that he raised these questions and was determined to confront them. He became the inspiration of his enthusiastic student, Lamarck, as well as another notable follower, Etienne Geoffroy Saint-Hilaire.

The idea of an affiliation of living beings that implied a variability of species was in good accord with Buffon's general philosophy. In his *First discourse on methods of study and interpretation of natural history,* he had not yet managed to detach himself from all the prevalent views held by the 'general public,' but it can be seen that he was already taking a very different approach in his studies of the procreation of animals. He saw continuity everywhere in nature; he could not even find a sharp demarcation between plants and animals.

'The general principles we set up are nothing but artifacts we devise to help us deal with great numbers of different objects in a systematic way, and, like the other artificial methods discussed earlier, they have the defect of never being adequate to encompass all things. This approach is inconsistent with the gradual processes and small, imperceptible stages by which nature normally proceeds. It is an attempt to explain an excessively large number of ideas with a single word the true meaning of which has become obscure. We try to sort out the natural world by saying that anything that is above a fixed line is *animal* and anything below it can only be *vegetable* or by drawing a sharp division between inert matter and organized life. As we have said more than once, such lines of demarcation do not exist in nature. There are creatures that are neither animal, vegetable, nor mineral, and it is futile to try to relate them to one another… We have said that nature proceeds by subtle degrees and often in imperceptible ways. The boundary between plants and animals is a very subtle one, but it tends to be more abrupt between plants and minerals.'[3]

In the case of plants and minerals, Buffon concluded that there were probably intermediate forms that were neither one nor the other. He had already pointed out that there are intermediaries of this kind between plants and animals: one of these is the fresh water hydra, a polyp of the water lentil [lemna] that had been the subject of memorable experiments by Trembley.[4]

Once one accepts the possibility of a general plan encompassing the entire animal kingdom with insensible transitions from one creature to another it becomes difficult to reconcile this view with the belief that everything in this world has a purpose. Buffon was strongly opposed to the doctrine of *final causes* that had dominated science since the time of Aristotle. He chose a very modest example, the organization of the foot of the pig, as a way of challenging the tyranny of this doctrine. He pointed out that although the foot has four digits, the animal makes use of only two of them. He wrote: 'The organization of the bodies of such creatures could not possibly be the product of some great plan. Why would Nature give an animal these useless parts, and at the same time leave it without other parts that would be far more useful?... Why is it so important that each individual part of the animal be useful and contribute to the body as a whole? Is it not enough just to observe that these parts are in harmony with each other and that each can develop independently without hindering the others? Any development that does not cause intolerable harm will survive if it is in harmony with the rest of the body… When we claim that all the parts of an organism contribute to an over-all plan, even if they have no apparent utility, we rationalize this by saying they must have some hidden purpose. We imagine relations that have no basis in fact and are not in the natural order of things but serve only to obscure it. We do not realize that we are altering philosophy when we distort its basic aim, which is to know the *how* of things and the way nature acts. Instead, we substitute the vain idea of trying to deduce the *why* of things, and this becomes an end in itself.

The questions raised by Buffon, a man who at one time was such a firm believer in the fixity of species, were the topics that would dominate thinking throughout the second half of the nineteenth century: the common origin of all living creatures, both plants or animals; the shared origins of animals of the same type; the populating of the continents by migration; the disappearance of ancient species that were victims of what Darwin later called the struggle for life; the appearance of new species arising from existing species that had been degraded or perfected; and the gradual evolution of humans. As his career drew to a close, Buffon became preoccupied by questions of this kind. All these ideas went through

---

[3] *Réflexions sur les experience de Leuwenhoeck.*

[4] *Histoire des animaux*, chap. II.

his powerful, logical mind, and they are precisely those that, with the aid of new research, are now beginning to triumph over all the older entrenched beliefs.

We live at a time when the limited means of observation force naturalists to seek out, somewhat reluctantly, more or less plausible hypotheses that can serve as provisional explanations of the inner-most phenomena of life and the mysteries of reproduction. Under such conditions, little could be added to what was already attempted by the ancient scholars. Buffon postulated indestructible *organic molecules* that come together briefly to form the individual plants and are then dissociated by the death of each individual and go on to become integral parts of other organisms. He was thinking along lines very similar to those of Anaxagoras. The organic molecules have nothing in common with the molecules making up the general character of the body. There are two distinct categories of matter, *dead* and *living*, and the two are not interchangeable. The living molecules are distributed everywhere, and when an animal takes in nourishment, it incorporates into its body only those organic molecules that match those of its own body and are appropriate to replace those it has lost.

'A living organism,' he says[5] 'is made up of similar organic parts just as we can think of a cube being composed of many smaller cubes. We see examples of this in our every-day experience. Just as a cubic crystal of salt is composed of many smaller cubes, we also see that an elm tree is composed of tiny elms. Whether it be the tip of a branch, a part of a root, wood from the trunk, or a seed pod, they all come from the same elm tree. It is the same with polyps and other species of animals that, when cut into separate pieces, can produce new individuals from each of the parts. And if our aim is to judge these things objectively, why should we try to find different interpretations for them?

'It seems very logical to me that, according to the reasoning we have just followed, there must exist in nature an infinite number of small organized beings, similar in every way to the large ones that make up the life of our planet. These tiny organized beings are composed of the living organic parts that are common to plants and animals, and they are elementary and incorruptible. As a result, reproduction or propagation is nothing more than a change of form by simple addition of these similar parts, just as the destruction of an organized being involves a division of these same parts… If we reflect on the way trees propagate and how such a small particle can reach a much greater size, we see that it must be by simple addition of small organized beings similar to each other and to the whole. A seed produces a tiny tree that it already contains in miniature. At the top of this small tree, a bud is formed that contains the small tree that will sprout from it in the following year, and the bud is still an organic part similar to the small tree of the previous year. At the tip of the second year's growth, an identical bud is formed for the growth added in the third year, and as the tree grows taller, it maintains the same form by putting out branches and budding miniature trees similar to those of the first year.'

---

[5] *Histoire des animauz*, chap. II.

Buffon's concept of plants differed little from the one Bonnet had reached; both expressed this idea in nearly the same terms. But Buffon was completely opposed to the idea that the miniature trees that were incorporated into the growth of larger ones had already developed their form while still in the seed. To explain procreation by this hypothesis of 'embodiment' of the seeds is to respond to a question with the same question. Buffon said that 'if we ask how one can picture the reproduction of living organisms, and the answer is that the new organism already existed, this is simply an admission that we are ignorant and have no hope of understanding it.' He said exactly the same thing about the hypothesis of immutable species. To respond to questions about the origin of species by saying that they have always been what they are now is to rule out any further research into their origins. From the scientific point of view, even a mistaken interpretation would be better than this discouraging doctrine.

In a similar way, Buffon also rejected any suggestion that the generation of species is now complete, and he opposed any hypotheses that calls upon ultimate causes, because, instead of proposing physical explanations, they appeal to arbitrary relationships and moral expediency. They lead to the famous hypothesis of *internal molds* that postulates that nature uses molds as a template to give organisms their external appearance and internal form.

The words 'internal mold' were poorly chosen, because a mold is normally used to reproduce a surface and not the internal details of a massive substance, but Buffon seems to have adopted the term for lack anything better. For him, all living beings have this internal mold, and certain forces penetrate the organic molecules causing the parts of the body to increase in size and weight without changing their form or structure. It is by means of this penetration of the organic molecules into the internal mold, and its compatible 'integration' that the living being develops, but the force that is responsible for the development is also that which governs reproduction.

All that is required is that some element of the living being be perfectly similar to the body as a whole, so that, when properly nourished, it may be capable of detaching itself and producing an identical, completely independent body in which it becomes an essential part.

'Thus, in the willow tree, just as in the polyp, there are organic parts that have all the elements of the body as a whole, and each piece of the willow or polyp that one cuts off from the main body can become another willow or a polyp.

Buffon added, 'an organized body of this kind in which all the parts are similar to the rest has a very simple composition and organization consisting entirely of identical parts repeated in a uniform way, and it is for this reason that the simplest bodies, the most imperfect species, are those that reproduce most readily. If an organized body contains only a few parts that resemble the body as a whole, reproduction will not be as easy or as abundant as it is in those in which all of the parts are similar. The organization of these bodies will be more complex than that of bodies in which all the parts are similar, because the entire body will be composed of parts that, though organic, are organized in a different way. The more complex the organization of this body, the more difficult it will be for it to reproduce.'

We find here the same ideas about organic perfection that we have already seen in Aristotle and which later led Milne-Edwards to propose his theory of the division of physiological work. Living creatures are continually using nutrition to add new molecules and new organic parts to their bodies, and a time comes when these new parts are over-abundant. They are then excreted from all parts of the body, including the testicles of the male and the ovaries of the female, to form the liquids that must be combined to produce a new living being. In the primitive forms of life, an unknown force causes organic molecules to penetrate the organs and bring about the growth of those that most resembled the molecules from which they were originally formed. Organic molecules representing the various organs of the individual will be found in the seed. The same force that makes them penetrate the corresponding organs will arrange them in the same order as they were in the parent. This theory of reproduction was published by Buffon in 1746. Maupertuis reiterated the same idea in 1751, but he endowed vaguely defined particles with mental faculties in order to avoid calling on the coordinating forces that Buffon had proposed.

In his theory of reproduction, Buffon did not perpetuate the older hypotheses, but, at the same time, he did not confine himself to pure reason. Whatever ideas he had grew out of observed facts. He said that 'we should always look for factual evidence to find new ideas.' He derived facts not only from observations but also from experiments. As Director of the Jardin des Plantes, he assembled collections of animals from all parts of the world and, wherever possible, observed them in the living state. The inability to reproduce is an unequivocal barrier that separates two species. Is there any way to cross that barrier? To what degree have hybrids developed into new species? Which are the wild species that man has used to develop his domesticated species? Buffon attacked all these questions with experiments. Time, he thought, was an indispensable element in resolving them, and he conceived the plan for an institution where these secular studies could be pursued. His plan was eventually realized in the form of the Museum of Natural History, which soon became the domain of lively scientific research and innovative thinking. Three great men came there to pursue Buffon's work: Lamarck, Geoffroy Saint-Hilaire, and Cuvier.

Chapter VIII

Lamarck

The importance of the lower animals – Spontaneous generation – Gradual perfection of organisms; the influence of needs and habits – Heredity and adaptation – Transformation of species during earlier geological periods – The absurdity of global cataclysms – Importance of ordinary every-day processes – Genealogy of the animal kingdom – The origin of man.

As a frequent visitor at Buffon's household and the traveling companion and guide of his patron's son, Lamarck can be considered the immediate successor of the illustrious philosopher and author of the *Epoques de la nature*. His style may not have been as effusive, but he had the same ability to digest facts and use them to develop illuminating concepts. He differed, however, in his scientific education and the focus of his studies. Buffon addressed much of his work to *physicists* rather than to *naturalists*, and combined an extensive knowledge of mathematics and physics with an unusual ability to deduce basic principles from general observations. His work had a precision and insight that is less apparent in Lamarck's. On the other hand, Lamarck's detailed studies of physiology of plants and lower animals enabled him to envisage broader relationships than anything Buffon was able to discern.

Those who devoted their work exclusively to the study of man and higher animals found it difficult to understand these complex forms of life and to offer explanations for the varied phenomena they display. To them, life appeared to consist of a host of different organs and functions appropriate to the nature of each individual. It seemed pointless, if not reckless, to attempt to penetrate the secrets of these forms of life and speculate on their origins. Lamarck had good reason to say that 'It is extraordinary that it did not become possible to explore the most important phenomena of life until we began to concentrate our studies on the most rudimentary animals and the complexities of their organization. It is equally remarkable that it was almost always by examining minute, inconspicuous objects of seemingly trivial importance that it was possible to gain the important insights that enabled us to discover the laws governing their development and trace its course through time.'

A. McBirney, S. Cook and G. Retallack, *The Philosophy of Zoology Before Darwin: A translated and annotated version of the French text by Edmond Perrier*,

It was considerations of the simple conditions under which life is manifested in the lower organisms that led Lamarck to conclude that these organisms were the first to appear, that they were produced spontaneously, and that all subsequent forms of life developed through gradual processes. 'Subtle fluids' activated by the heat and light of the Sun penetrated tiny particles of mucilaginous, inert matter that were receptive to their action, and by animating them, produced the first living beings. He believed that these fluids are still capable of animating inert matter and that new organisms, such as the infusoria, are continually being formed by this process of *spontaneous generation.* After Lamarck made this proposal, it had the effect of relating the gradual evolution of life more firmly to spontaneous generation, even though there was no compelling reason for integrating these two concepts. At a certain moment in the evolution of the Earth, the prevailing conditions were said to have activated certain substances with the unique attributes of life and enabled them to create new forms of life by passing on these attributes in differing ways to other types of inert matter. There is no evidence that these conditions still prevail. The extensive experimental studies of Louis Pasteur have long since ruled out any possibility of spontaneous generation in the inert realms of our environment.

As for the origin of primitive organisms, Lamarck simply said, in the limited language of his time, that it could be accomplished only by activating material with a special kind of mobility, and that the molecular movements of the very simple organisms produced in this way have gradually been perfected by the persistent action of these same subtle fluids. Lamarck supposed, as did Erasmus Darwin, that these organisms resulted from new stimulants, new *needs*, that multiplied in each living being at the same time that its organization became increasingly complex and its relations to the external world more diversified.[1] But while his English rival believed that the agitation produced in the organism by these new needs was sufficient to generate new organs or modify existing ones, Lamarck introduced an intermediate stage between the appearance of new needs and the physical modifications that resulted from them. In his view, these persistent needs have led to a ceaseless repetition of certain acts and a development of certain habits that in turn gave an additional impetus to the modifications. In effect, the animal's organs are developed and perfected by their frequent, habitual use. In contrast, organs that an animal ceases to use will atrophy and disappear. Thus, habitual use can cause certain organs to be replaced by others that are gradually perfected. There is no question, for example, that the eyes of animals that habitually live in darkness tend to disappear, and common, every-day experience leaves little doubt that most organs are developed by exercise. But this process of diversification assumes that such organs already exist. How can an entire new

[1] There was an extraordinary similarity between the timing, wording, and illustrative examples used by Erasmus Darwin and Lamarck when they both proposed these same ideas almost simultaneously. After weighing the evidence, Osborn (1894, p. 152–155) concluded that no plagiarism was involved. It was a pure coincidence, largely the result of the strong influence that the work of Buffon had on both men. (Trans. note)

organ be constituted from nothing? Lamarck goes beyond the valid implications of his hypothesis when he supposes that the animal's need for an organ is in itself sufficient to make it appear. It is difficult to explain, for example, how the ruminant animals acquired horns by saying that 'the surges of anger that males frequently experience induced a stronger flow of fluids toward the appropriate part of their head where they caused horn-forming secretions that mixed with bone-forming material to produce a solid protuberance.'

Ruminant animals are not the only example to which Lamarck applied his theory that exceptional activity generates an *internal effort* that, in turn, directs fluids to the parts of the body where they are needed. 'When the animal's will causes it to act in a certain way, the organs that must execute that action are immediately stimulated by the activating fluids that produce the desired action... Repeated stimulation and use of this system will not only enhance the development of the necessary organs but may even *create* them.' This amounts to saying that an animal comes to possess whatever organ it requires or finds useful in the ecological conditions in which it has been placed. Lamarck has been severely criticized for this statement, which was, it is true, a bit rash. It has been maliciously misquoted as: 'An animal always ends up possessing the organs it wants.' This was not what Lamarck had in mind. He simply attributed the transformations of species to stimulation by the external conditions to which an organism needs to adjust, and this was his way of explaining what we now call *adaptation*. Thus, the long neck of the giraffe results from the animal living in a land where the leaves are concentrated near the tops of tall trunks. The long legs of wading birds result from these birds having to search for their food without submerging themselves in the water, and so on. These interpretations do not detract in any sense from the two basic laws that Lamarck set down:

'1. In all animals that are still in a stage of development, the more frequent and sustained use of any organ strengthens, develops, and enlarges that organ and gives it strength in proportion to the duration of this use, while a lack of use of such an organ gradually weakens it and causes its utility to deteriorate progressively until it finally disappears.

'2. All the features that the individual has acquired or lost through the influence of the natural conditions to which a race has been exposed for a prolonged time and through the influence of the frequent use or lack of use of that organ, are passed on to descendents provided that the acquired changes are shared by both sexes that produce the new individuals.'

Today we can add numerous other examples to those that Lamarck assembled in support of the first of these laws. The only point that might call for discussion in this regard is the extent of the changes that an organ can undergo through the use the animal makes of it. This is simply a matter of degree. The possibility of creating an organ through external stimulations is itself a point that merits study. One does not have the right to treat it as a ridiculous fantasy and reject it without careful, objective studies. Lamarck's ideas would no doubt have gained wider acceptance had he not chosen to call on the intermediary role of need. There is no question that, lacking stimulation, organs begin to atrophy and eventually disappear. We have already mentioned that creatures in dark caverns or at great depth in the sea are commonly blind. The proteus living in the underground lakes of the Caroline Islands is white, and under the action of light its tegument develops pigment and turns brown. Light is unquestionably necessary for the formation of chlorophyll in plants. In both cases, whatever the mechanism by which the pigment or chlorophyll are formed, they appear only in response to an external stimulus.

Lamarck's concept of life was closely tied to his hypothesis about the way organs form and develop, and his hypothesis, when considered from this point of view, loses any unreasonable appearance it may have had. It commands respect as an unsuccessful attempt on the part of a great mind exploring all the knowledge acquired until his time to find a solution to a problem that, despite all the progress that has since been made, we have still not been able to make nature reveal.

According to Lamarck, two forms of energy permeate the molecules associated with life: *heat* and *electricity*. Heat causes living molecules to distend and distance themselves from one another without destroying their cohesion and in this way keeps the living tissues in a special state of tension that Lamarck referred to as *orgasm*. This orgasm, which is essentially a struggle between the cohesion of the living molecules and the effects of heat, enhances the *sensitivity* of the tissues. Thus, if electricity, always in motion, happens to influence some point, either from external influences or through volition, the equilibrium between heat and cohesion at this point is destroyed, and the tension [orgasme] ceases. The tissue, which is no longer in a state of tension, contracts over the place where the heat was weakened, and quickly resumes its original state. In this way, the tissue reacts to external stimulations. A muscle that is not contracted manifests its state of tension by what we call muscular *tone*. Do the muscles and nerves serve as instruments of the will by transmitting the electrical impulse that brings a spasm to an end and allows the muscles to contract and return to their normal state? Today we would no doubt have other explanations for all the phenomena that Lamarck attributed to this spasm, but do we really know that much more about the origins of life? When we say that one must think of it as motion of protoplasmic particles, motion that we are unable to define, do we express an idea that is basically different from Lamarck's, since, by definition, heat is nothing but a form of motion?

Has our search for the causes of the modifications of the organisms been any more successful? Even if we no longer agree that the needs and desires that provoke these changes could by themselves cause the appearance of new organs or major changes in existing ones, we can hardly question the effects of their use or disuse; one can no longer doubt the direct effects of the environment. We accept the precept that when one organ is modified, several others must respond by adjusting to the change, either by developing in accord with it or by reducing any effect they may have on it. There is much evidence that indicates that the rapidity with which developing organs become more complex and firmly established can also induce changes in the way related parts of the body interact with them. We also accept a certain amount of spontaneity in these organic variations. In some cases, interbreeding may be an important factor, but modifications introduced in this way are not strictly new but only inherited characteristics that were produced by the same causes we have just enumerated. Until now, there have been no systematic studies of the influence of external conditions and the various modifications they can produce. Even Darwin limited himself to stating that species vary without asking why. This simple fact serves as the basic premise of the theory of natural selection, but we must defer a closer examination of this precept until later.

Lamarck's second law, the one for inheritance of characteristics, was a keystone of Darwin's edifice. He recognized that the struggle for life has the inevitable effect of eliminating static forms and useless variations so that the more advantageous ones could survive. This explained why there is so little continuity between all the contemporaneous forms living side by side. It also explains why those that have survived may appear to have degenerated from earlier forms but have actually become better adapted to the ecological conditions in which they live. They display such marvelous harmony with their environment that one could easily believe they were specially created in the manner proposed by the theory of *ultimate causes.*

Like Buffon, Lamarck was adamantly opposed to the Aristotelian doctrine of ultimate causes. Far from believing that living species were created *for* a particular kind of life, he affirmed that they are created *by* the kind of life imposed by the conditions in which they find themselves. For him, the marvelous adaptation of species is proof of the direct action of the environment. His theory of transformation, instead of explaining this adaptation as Darwin did, accepted it as a well-established precept. The differences between the methods of the two great naturalists are worth noting.

Species, because they are adapted to the conditions in which they live, must remain immutable so long as those conditions remain unchanged. This was Lamarck's response to an objection that some believed would overturn his entire system along with that of Darwin as well. During the expedition to Egypt, Geoffroy Saint-Hilaire visited the necropolis and collected a great number of mummified animals that he and Cuvier studied on his return to France. These animals that had died several thousand years earlier were found to be identical to animals living in Egypt today. Cuvier believed that this was clear proof of the immutability of species. At that time, the duration of geological time was poorly known. Instead of periods of millions of years that the geologists now assign to our world, many believed that the time span of creation was scarcely six thousand years and that the mummies of the hypogeum of Egypt could be representatives of the earliest ages of the Earth. We now know, of course, that their great antiquity is only an illusion and that nothing, not even man, has changed significantly. The span of time separating us from the period when the mummified animals lived was as brief as a flash of lightening compared to the time required for nature to produce a new age.

As others have quite properly pointed out, the fact that the forms of mummies are identical to those of modern beings is sufficient proof that conditions have not changed in this relatively brief period. It is not just the natural species that have remained unchanged, but also the domesticated animals, which we know have been altered to a much greater degree by selective breeding.

Having already studied the great varieties of fossil molluscs that can be traced through successive variations more easily than mammals, Lamarck found numerous forms that were transitional between supposedly extinct species and those that are living today. He concluded that species do not become extinct but only change their form.

He wrote that,[2] 'if there are species that have in fact disappeared, they must have been among the large animals that lived on dry land where man exercises an absolute dominion and has been able to destroy all the individuals of species that he did not consider worth preserving or domesticating. This raised the possibility that, although animals of Cuvier's genera *Palaeotherium, Anoplotherium, Megalonyx, Mastodon* and a few other species of known genera may no longer exist in nature, *one cannot rule out the possibility that they still do.*

'But animal species that live in the water, especially marine waters, as well as small, inconspicuous races that inhabit the land surface, are less vulnerable to man's assault. Their rates of reproduction are so high and their means of avoiding his pursuits and traps are such that it seems unlikely that he could destroy an entire species of these animals.'

Like Buffon, Lamarck was impressed by the importance of man's role in nature and considered him the sole cause for the disappearance of species. He did not point out that the war that humans have declared on animals is only one example of the great struggle that goes on among the animals themselves, but he did not fail to recognize the main consequences of that struggle, for he wrote:[3]

'With the multiplication of weak species, and especially the more poorly adapted animals, the resulting competition of great numbers of individuals could limit the survivors to those that had acquired the most effective adaptation. In essence, nature has not constrained the size of an increased population to the numbers it can reasonably sustain.

'Most animals eat one another, and even those that live on vegetation may be eaten by carnivorous predators.

'It is well known that the strongest and best armed eat the weak, and that the large species devour the small.'

What is proposed here seems to be quite close, not only to the struggle for life as envisaged by Darwin but even to natural selection. Unfortunately, instead of pursuing the idea, Lamarck chose to follow another path. He does not seem to have appreciated the consequences of the intense competition that arises between

---

[2] *Philosophie zoologique,* 1809 vol. I, p. 76.

[3] *Philosophie zoologique*, vol. I, p. 98.

animals of the same species whenever the available food supply is insufficient. On the contrary, he believed 'that the individuals of a given race rarely eat one another or have conflicts with other races.' He then unconsciously fell back on final causes when he developed the idea that nature has taken precautions to prevent large species from multiplying to the point that they would become a threat to the existence of smaller ones. Darwin took a view that was exactly the opposite of Lamarck's, but one cannot accuse the latter of failing to address a problem that was not even stated at his time, namely that of the gradual extinction and renewal, independently of human influence, of most species of plants and animals.

Although Cuvier was a firm believer in the fixity of species, he did not hesitate to affirm that numerous animals had disappeared over time, and we shall see that he attributed their disappearance to great catastrophes and general cataclysms that overwhelmed the Earth. Lamarck, on the other hand, was impressed by the gradual transformations he had found in molluscs and questioned the reality of these global revolutions which Sir Charles Lyell and his disciples would later show to be absurd.

'Why,' he asked,[4] 'should one propose a *universal catastrophe* for which we have no tangible evidence when better-known natural processes offer more reasonable explanations for the observed facts? Nature never acts suddenly; it always proceeds slowly by successive degrees, and any disorder we may see is the result of local, isolated disturbances of the normal regime. These principles offer rational explanations for everything we observe, and there is no need to call upon universal catastrophes that disrupt large parts of the natural regime.'

The most reasonable doctrine is one based on common, every-day causes. The principle that geological features can be explained by the same slow processes we see operating today has long served as a reliable guide for all the major schools of geologists.

When he considered the various classification systems in terms of his theory of evolution, it seems that Lamarck looked back to Bonnet's scale of living forms and quickly saw that there is no simple way of dealing with animals as a single linear series. He divided animal life into two lines each of which had a progenitor created by spontaneous generation. Some of these animals are born fully developed and self-sufficient, while more highly organized animals begin life within the body of a parent and remain dependent on the adult through a long period of development. The latter existed first as parasites and make up the class of helminthes [worms]. The first type evolved to a very limited extent, while the second culminated in the vertebrates. Instead of placing the latter at the head of an animal kingdom, Lamarck was the first to proceed from the simple to the complex by starting with the infusoria or simplest helminthes [worms] and gradually rising to the most perfect forms that life attains.

---

[4] *Philosophie zoologique*, vol. I, p. 80.

'The order of nature,' he said, 'is the same as that in which individual bodies have developed from the first spark of life,' and, as all these bodies seem to have come from earlier, simpler forms, it is evident that the members of any more or less comprehensive group must have formed uninterrupted series with no clear demarcation separating one individual from another. 'Nature has not created distinct classes, orders, families, groups, or species but only individuals that succeed one another and resemble both their predecessors and successors.' Those individuals that share a strong resemblance and retain their identity from generation to generation constitute a *species*. But the individuals making up the species preserve constant characteristics only if the circumstances in which they are placed also remain constant. As soon as conditions begin to change, the individuals also begin to change. A certain number of intermediate varieties grade into one another to form links between animals of very distinct forms. Thus, there is no such thing as an invariable species.

Lamarck exaggerated the number of these intermediate forms that can exist between species.[5] He also exaggerated the facility with which the species can be crossed, for he thought the species were too unstable, but he was not yet aware of the important fact that species disappear, and at that time it seemed impossible that there could be such a void in nature. Lamarck refused to believe that the gradations between species are firmly fixed, as others had sometimes supposed. He saw a profound *hiatus* between the simple and more highly organized bodies,[6] and he believed that a similar hiatus separates plants from animals. Animals, he said, possess a faculty, which he referred to as *irritability*, that was entirely absent from plants. The plants and animals within a given class have a definite scale of development in which their organs range from the most rudimentary to the increasingly complex. This scale represents 'the inherent order of nature and results, along with all else that this order entails, from the means with which the supreme Creator of all things has endowed them. It is part of the larger, immutable order that this sublime Creator established and in which everything is governed by the general and particular laws of nature. By these means, which it continues to employ in a constant way, the Creator has given and will always continue to give its creations an existence. It changes and renews them endlessly and in this way maintains the comprehensive order that is its principal effect.'[7]

For Lamarck, the various forms of plant and animal life result from one or the other of two possible causes:

1. A certain natural order, which was directly instituted by the Creator, and manifests itself in the unique and subtly graded scale of development seen in all plants and animals.

---

[5] *Philosophie zoologique*, vol. I, p. 58.

[6] *Philosophie zoologique*, vol. I, p. 92.

[7] *Philosophie zoologique*, vol. I, p 113.

2. The influence of external conditions that, without altering this order in any essential way, produces a multitude of natural variations and creates within the unique scale that distinguishes each realm of life an infinite number of small secondary series some of the branches of which may appear to be completely isolated.

This is an important argument. Lamarck is often represented as having exclusively attributed the evolution of the universe to natural forces. Hæckel, in his *Histoire de la creation Naturelle*[8] repeated this opinion. This was not, however, the thinking of the author of the *Philosophie zoologique*. Of course, it was Lamarck's belief that the material and 'subtle fluids' that constitute what we now call physical-chemical forces, were sufficient to form the simplest of the living beings. External conditions have no doubt played a major role in the origins of the organized forms of life, but their further development into a gradational series of increasingly complex organisms followed a plan set down in advance by the 'supreme author of all things.' It seems that Lamarck was attempting to find a way to incorporate environmental effects into Bonnet's idea that there is a gradational scale of living forms. It would have been hard for him to reject that idea completely, because it must have seemed to be in accord with his own conception of the majesty of the Creator. There were definitely occasions when Lamarck felt compelled to think in terms of final causes, and this led him to say:[9] 'Thus, *this wise foresight* maintains an *established order* in all things. The changes and perpetual renewal that are observed in this order have definite limits that cannot be exceeded. The variations that all creatures undergo do not prevent them from subsisting. The progress creatures make in perfecting their organization is never lost. All that appears to be disorderly or anomalous eventually falls into place and even contributes to the general order. *Everywhere, the will of the supreme Author of nature and all that exists in it is invariably executed.'*

One could hardly find a better way to express the theory of final causes, for if God has arranged and coordinated everything so that his will may always be fulfilled, then everything has been planned and, as the creationists would have it, all the powers with which he has endowed nature are designed to conduct it toward an inevitable foreordained end: the achievement of the creator's will.

There was, however, an astonishing contradiction in Lamarck's thinking. His view of nature as a whole was that of a creationist, but when it came to details he was a resolute adversary of final causes. The works of naturalists and philosophers are filled with admiration for the marvelous tools with which animals are provided and the wonderful ways in which these tools are designed to serve a specific function. For most of them, it is indisputable proof of the intelligence and wisdom that presided over creation.

---

[8] *Histoire de la creation naturelle* French translation by Reinwald, 1874, p. 102.

[9] *Philosophie zoologique*, vol. I, p. 101.

'The fact is,' said Lamarck,[10] 'that all animals have patterns of behavior that are characteristic of their genus and species, and the design of their bodies is in perfect accord with their behavior.

'On considering this fact, it seems that one is led to one or the other of the two following conclusions, neither of which can be proved.

'*The conclusion accepted until now:* When nature (or its Author) created the animals, it foresaw all the possible conditions in which they would be living and gave each species a definite, fixed form that forced it to live in the places and climates where it is found and to preserve its established habit.

'*My own personal conclusion:* Nature, in producing all the successive species of animals starting with the simplest and most imperfect and ending with the most perfect, has gradually elaborated their organization and enabled them to spread into every corner of the habitable world. Each species has been influenced by the conditions in which it finds itself and has adopted the behavior and the modifications of its body that we see in them today.'

Lamarck had no hesitation in choosing between these two conclusions. The first assumed that the species are fixed and have always been as well adapted to living conditions as they are today, but this fixity of species assumes in turn that the conditions imposed on the species are also fixed. But these species no longer resemble the animals from which they have descended, and we can continue to modify them at will. The argument that conditions have remained fixed is clearly contrary to all we know about the geological past. Moreover, we have deliberately changed the living conditions of a certain number of animals, principally the domesticated ones, and there is no question that animals have been modified in response to the new conditions. None of them resemble the wild animals from which they have descended, and we can continue to modify them at will. The argument is irresistible; it is a solid obstacle to all reasoning in favor of the fixity of species.

What these arguments come down to is this: the modifications imposed on domestic animals have had certain limits. One could respond that until now no one has tried to change an animal's original form in any major way; man has only developed desirable features that already existed. He has not attempted to transform the animal by introducing profound changes; he wants only to conserve and perfect some trait that seems advantageous. Even if he could create something new, it would require much more than the six thousand years that have elapsed since man first began to domesticate animals. The work of nature takes millions of years; why should we be surprised that man has not yet been able to overturned it?

Lamarck accepted without reservations the view that the present species have evolved through modifications of earlier ones. He believed that infusoria, born directly by spontaneous generation, have perfected themselves to produce radiolaria.

---

[10] *Philosophie zoologique*, vol. I, p. 265.

Worms that are formed inside the bodies of other plants or animals have had a more rapid evolution. They are divided into two branches, one of which has led first to insects, then arachnids, then to crustaceans. The other has produced a succession starting with the annelids and proceeding through the barnacles, molluscs, fish, and reptiles. In the latest bifurcation, reptiles lead on the one side to birds, which in turn gave birth to monotreme mammals, and on the other, reptiles produced the amphibian mammals and the latter were the source from which were derived first the cetaceans then the ordinary mammals that finally divided themselves into the ungulates [animals with hooves] and non-ungulates [animals with claws or nails]. The genealogical table of the animal kingdom shown here was the first to be based on this kind of scientific evidence.

Many documents that could now serve to construct such a tree of life were not available to Lamarck, so it is not surprising that he reversed the order in which the arthropods evolved, that he mistakenly interpolated the barnacles, which are crustaceans, between the annelids and molluscs, that he had the monotremes descending from birds instead of joining them with the other mammals, and finally, that he tried to derive ordinary mammals from the amphibians instead of having these animals descend directly from reptiles, as we do today. Reversals like these were inevitable as long as knowledge of their genesis was incomplete, and there have been other examples of this since Lamarck's time, but they are becoming less common as science continues to progress. The essential thing was his recognition of a relationship between the different types of organisms that has since been largely confirmed.

It will be noticed that man is not included in this table. Lamarck's thoughts regarding the origin of humans have been presented in various ways, and it is best to cite his own words: 'If the only thing that distinguished man from the animals were his physiological form, it would be easy to show that the characteristic features of the human race, with all its variety, are the result of ancient changes in the physical conditions in which he lives and that the behavioral patterns he has developed to adapt to these conditions have become distinguishing marks of his species.'[11]

Lamarck shows how a well-developed race of animals with four limbs, if it ceased to climb, could become two-legged and how it could acquire an upright posture through the need to ensure its security and explore distant parts of it habitat. By collaborating with other members of its race he could dominate rival species and confine them to the forest, and the need to communicate with other members of his group would have led to the development of language.

[11] *Philosophie zoologique*. vol. I, p. 349.

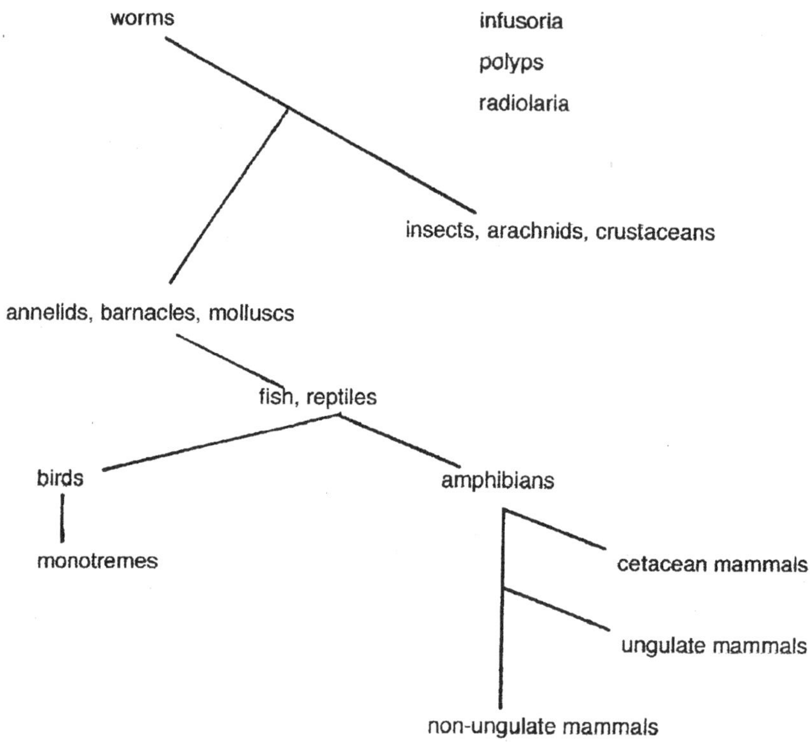

Fig. 8.1 Diagram showing the genetic relations of the different animals

'Thus,' he added, 'the response to need can account for everything we see in the modification of species. It has directed the development of the organs needed for the habitual performance of physical acts and even the articulation of sounds.

'Such would be the reasoning one could follow if man, considered here to be the pre-eminent race in question, were distinguished from animals only by the characteristics of his physical organization and if his origin *were not different from theirs.*'[12]

This opinion can be summarized as follows: as a naturalist, Lamarck did not hesitate to consider man as a modified primate, but as a philosopher and psychologist, he saw such a chasm separating man and the animals that he could only conclude that man has emanated directly from the will of the Creator. This reasoning would be sufficient to make the concept of transformism acceptable to many of those who now dominate respectable religious sects. But what interest could the doctrine of evolution have if it ends precisely at the point that is most important for us to elucidate, and if, after claiming that it has revealed the origin of all the animals, it leaves us completely ignorant of the prior history of our own species?

---

[12] *Philosophie zoologique*, vol. I, p. 357.

And yet, even from the psychological point of view, the barrier that Lamarck established between man and animals was a quite feeble one. It will be recalled that, according to Lamarck's beliefs, external conditions do not act directly on the organisms; they only modify them by stimulating a response to the new habitual use or disuse of the organs, and in this way they determine whether these organs develop or atrophy. The sensations created by new needs are the link by which the need for a response is transmitted to the animal's mental faculties. Lamarck attached great importance to the degree to which these faculties are developed in the three levels of animals he defined as *apathetic, sensitive*, and *intelligent.*[13] The simple statement of this classification is enough to show that Lamarck accepted a gradual development of the mental faculties. He also endeavored to demonstrate that 'all acts of judgment require a system of special organs in order to carry them out,' and as these organs are the same in man as in the higher animals, there is only a difference of degree between them. It must therefore follow that if the highest animals come from simpler ones, man in turn must come from the highest members of the animal kingdom. After having developed his ideas about the nature of mental processes, which he regarded as simply an assemblage of mechanical phenomena, Lamarck could have applied these concepts to the problem of man's place in nature, but he did not.

One wonders whether he may not have feared that by taking this last bold step he might compromise the success of all the work to which he had devoted an incredible intellectual effort and that he knew to be much in advance of the time. He closed his book with a melancholy reflection that, unfortunately, is still true: 'Men who endeavor to push back the limits of human knowledge realize that it is not enough for them to discover and demonstrate a truth that is new and useful; they still have to disseminate it and make it widely known. An *individual's reasoning* in assessing the validity of something new is not necessarily the same as that of the *general public*, and this often makes it more difficult to gain recognition of a truth than to discover it. I leave this subject without pursuing it further, knowing that my readers will be able to make allowances for any deficiencies, provided they have had the experience of observing the motives that determine men's actions.'

This simple, impartial sentence expressing his warm-hearted philosophy reflects nothing less than a clear expression of how Lamarck felt about his contemporaries. One of them has left this anonymous comment on the copy of *Philosophie zoologique* in the library of the Museum: '*what a superficial man.*' This outspoken reader expressed rather well the impression of those who did not appreciate the contributions of a great naturalist who dared to envisage the entire organic world from a different point of view. Lamarck had thrust himself upon zoologists by publishing *Histoire naturelle des animaux sans vertebras* and thereby earned the title of a French Linnaeus. As Isidore Geoffroy Saint-Hilaire said, one can pardon his zoological philosophy because of his great descriptive contributions.

---

[13] *Histoire naturelle des animaux sans vertebras.*

But the innovative, highly fruitful ideas that he so generously expressed throughout his book were soon buried under sarcasm. It is regrettable that even Cuvier was associated with this slander. Lamarck's innovative thoughts should have been allowed to sleep for half a century before being exposed to the meditations of other scientists.

The man who first tried to identify in precise scientific terms the genealogical ties that unite the simplest animals to the more highly perfected ones and was also the first to penetrate the importance of heredity, dared to say that we must look to the past for explanations of the natural creatures living today. He also proposed the general rule governing the development of life on our planet by the slow, gradual evolution of organisms without sudden leaps or cataclysms. The man who was the first to try to explore the mysteries of life in the light of the physical sciences will always deserve universal admiration. It is true that the real mechanism of perfecting organisms escaped him, but Darwin did not explain it any better. The law of natural selection does not indicate a process by which animals are transformed; it simply describes the results of such a process. It states these results without showing us how they have been achieved. We see quite clearly that it leads to the survival of the most perfected organisms, but Darwin did not show us how these organism originated. It is only in recent years that anyone has tried to surmount this deficiency.

Perhaps Lamarck's ideas could have won their due recognition sooner if they had not been developed at a time when the scientific arena was almost entirely occupied by two younger, more ardent contestants: Geoffroy Saint-Hilaire and Georges Cuvier.[14] We must not omit from this historical sketch these two names that are so often linked in the academic debates of the first half of this century, names that remain inscribed on the banners of the two rival schools and can be considered the most eloquent champions of two opposed turns of the human mind.

---

[14] Osborn (1894, p. 179–180) remarked that Lamarck's use of overly speculative examples to support his theory was largely responsible for his failure to gain a wide acceptance of his ideas, but more recent research has tended to lend support to some of his seemingly outlandish ideas. For example, he postulated that snakes were reptiles that had lost their legs through lack of use, but we now know that there are indeed reptiles that have only vestigial legs because they live under conditions where the legs are an unneeded encumbrance.

Chapter IX

Etienne Geoffroy Saint-Hilaire

The opposed doctrines of fixity and variability of species. – A unified plan of composition – Importance of the rudimentary organs – Balancing the organs – Analogy of the lower animals to the embryos of higher animals – Arrested development – Monsters and teratology – Geoffroy's ideas about the variability of species; abrupt transitions; influence of the environment – Extension of the theory of the unified plan of composition of articulated animals: return of the vertebrates; the ideas of Ampère – Genetic links between fossils and living species.

Two opposed opinions regarding species became well established in science, and each soon had its ardent supporters. Linnaeus held to the absolute immutability of species, while Buffon, and especially Lamarck, insisted on their variability. According to the latter, a species is capable of undergoing limitless modification. Buffon did not assign great importance to this concept, but Lamarck thought that genetic modifications were essentially unlimited because, in his view, the most advanced species have descended from simpler ones through an uninterrupted series of generations. This same conceptual difference would later be found in the works of Cuvier and Geoffroy Saint-Hilaire whose ideas were explored within the more restricted circles of the Jardin des Plantes and the Academy of Sciences of Paris. Two minds, both of the highest order, became engaged in one of the most famous intellectual struggles in the history of science. Geoffroy Saint-Hilaire's home ground was the king's garden which Buffon had made so famous. It was there that he carried on his scientific studies and received his training in anatomy under the influence of the author of the *Histoire naturalle*, Louis Daubenton. Daubenton had been one of Buffon's faithful collaborators and replaced Lacépède as assistant curator and demonstrator in the department of natural history. In 1793, a decree of the national Convention established the *Muséum d'histoire naturelle,* and Geoffroy was assigned to the position of mineralogist. Scarcely twenty-one years old, he had the title of professor of zoology in the new 'capital of natural sciences.' One of his duties was to teach the natural history of vertebrates while Lamarck was charged with that of the invertebrates. From then on, the scope of the young naturalist's studies became more sharply focused. This was a time when the vertebrates were still considered the most important of all animal life and were thought to typify the animal kingdom as a whole. Geoffroy undertook an impassioned study of their organization and became intrigued by the many basic

A. McBirney, S. Cook and G. Retallack, *The Philosophy of Zoology Before Darwin: A translated and annotated version of the French text by Edmond Perrier*,

resemblances he found among them. Buffon had noted that nature seemed to follow a regular pattern 'from humans to quadrupeds, from quadrupeds to cetaceans, from cetaceans to birds, from birds to reptiles, from reptiles to fish,' and Geoffroy set out to demonstrate that this pattern was indeed real and endeavored to determine its exact relationships.

For one who was fascinated by this long series of organisms ranging from humans to fish, it must have seemed that nothing could be a more timely subject of research. Geoffroy soon concluded that the same general pattern seen in the relationships of the forms of life he was studying could be found throughout all nature. Starting in 1795, at the age of only twenty-three, he had a very close association with Cuvier, whom he had just brought to the Museum of Natural History. He wrote in his *Mémoire sur les rapports naturels des Makis*; 'Nature has formed all living beings according to a unique plan. Although the details of their relationships vary in a thousand different ways, they are all based on the same fundamental principles. This plan becomes most apparent when we consider a particular class of animals and observe how it has enabled various forms to develop from one another. By simply changing the forms of some of their organs animals can adapt them to new functions and extend or reduce their use.... All the most important differences that affect each family of a given class result from different arrangements or modification of the same organs.'

Buffon had pointed out that a great many animals are constructed on the same plan, and Geoffroy later affirmed that all animals did indeed have the same fundamental structure. This idea of the *unified plan of construction* of animals, so grand and yet so simple, would henceforth be the central theme of almost all his work, and demonstrating it would become his main concern. In his studies of animals he was not searching for differences between them, as the disciples of Linnaeus were, but for resemblances that they might share, and by 1796 this preoccupation had already led him to an interesting result. In summarizing the conclusions of his research on marsupials, he pointed out the resemblances between the dasyures and civet-cats, phalangers and squirrels, kangaroos and jumping-mice (jerboa), and in this way, he established a kind of parallelism between the marsupials and other mammals. This was the first time that anyone had proposed the idea of *parallel classifications.* His son, Isidore Geoffroy, would later develop this idea and demonstrate its importance.

In Geoffroy's view, however, 'there are things in natural history that are more important than classifications.' The principal one was genetic relationships, a study he pursued enthusiastically in the belief that it would explain many types of natural phenomena. The idea that all forms of life have a common line of descent naturally drew him to the work of Charles Bonnet and his 'ladder of life', but his knowledge of zoology told him that he could not stop there. In 1794 he said that 'this universal chain is nothing but fantasy.' But he knew quite well that living beings are not isolated from one another and that, despite their diversity, they are closely linked together. He wanted to find a better formulation of Bonnet's hypothesis and believed that he had found that the unified plan of construction was the basic law of nature. It should be noted that this idea, which became Geoffroy's crowning contribution and the inspiration of all his studies, had led him to discover the principles that have dominated the work of naturalists of even the most opposed schools. This seminal idea was not the culmination of his long career of tireless investigations, even though it had been in the back of his mind since he was very young and just beginning his studies. Fundamental ideas do not spring from a mind that has begun to decline after spending years exploring a maze of factual minutia. Why would the benevolent muses choose to illuminate the last work of one who, throughout his entire life, has held them in disdain? They often behave capriciously, making themselves elusive for many years only to return and shine a playful light into a complacent mind that tries to pass them off as an ethereal dream. They may become tired of this resistance and gradually withdraw until the sweet vision vanishes, never to return. Those who fail to capture that vision are left with nothing but the pleasant memory of a broken charm, and yet, these alluring phantoms are the very driving force of the human mind. They are responsible for opening new pathways for genius to follow and for showing the route that will open new horizons. Although they demand the total loyalty of those who have benefited from their gifts and must be accepted whole-heartedly, that is a small price to pay if those who follow them are able to reap a rich harvest for the benefit of humanity.

Such was the case for Geoffroy Saint-Hilaire. He dreamed of finding an explanation for the marked resemblances of animals, and he thought he saw the solution in a unified plan of composition. The muse did not allow him to settle for half a solution. It had already given hints to Aristotle, Galen, Ambroise Paré, Belon, Newton,[1] Vicq-d'Azyr,[2] Buffon, Goethe, Herder, and Pinel, but it was only Geoffroy who had enough perseverance to uncover her precious secrets.

During the expedition to Egypt, observations of the wings of the ostrich made him conscious of the potential importance of rudimentary organs. He noted that the breast bone of this bird is very poorly developed, and remarked that 'these rudiments of the breast bone have not been completely suppressed because nature never moves by sudden changes. Even when it is entirely superfluous, vestiges of an organ will remain if it has played an important role in other species of the same family. Thus vestiges of the wings of the cassowary can be found under the skin on its flanks, and we see in humans a bloated flesh at the internal angle of the eye that is recognizable as the rudiment of a membrane that many quadrupeds and birds still possess.'

Around this same time, in 1800, he wrote: 'The seeds of all organs that are seen, for example, in the different families of animals that breath with their lungs,

[1] We read in Query 31 of Newton's Optics: 'And so must be the uniformity in the bodies of animals; they have a right and a left side shaped alike, and on either side of their bodies two legs behind and either two arms or two wings before upon their shoulders and between their shoulders.

[2] See *Vie, travail et doctrine d'Etienne Geoffroy Saint-Hilaire* by Isadore Geoffroy Saint-Hilaire p. 143.

are present in all species. The cause of the infinite diversity of forms that distinguish species in which degenerated or completely atrophied organs are preserved, must be related to the proportionally greater degree of development that some of them have, *a development that always operated at the expense of other organs in neighboring parts of the body.'* This latter comment was the first suggestion of what Geoffroy Saint-Hilaire would later call the *principle of balancing of the organs.* This principle offers an explanation for the preservation of rudimentary organs as incomplete products of embryonic cells that have been aborted because other organs had deprived them of the nourishment that was originally intended for them.

It is rare for organs to disappear completely. The rudiments, though imperfect, still occupy the same parts of the body as the more fully developed organs to which they correspond. This important fact demonstrates a fundamental principle of the unified plan of composition.

As we have seen, a similar unity assures that *all animals of the same group* possess the same organs. (Geoffroy seems to have moderated the absolute statement he had made about this in his memoir on the makis.) But how can one recognize these corresponding organs in a series of innumerable related forms? Geoffroy seems to have visualized a method of investigation based on his belief in a unified plan of composition and applicable to all animals constructed according to the same plan regardless of how many such plans nature may follow, and, under the name *theory of analogs,* it would provide anatomists of all schools with one of their most useful tools for exploring unknown realms. One can think of the organs in various ways, notably from the point of view of their physical form, their function, or their relative position with respect to the body as a whole. When the organs of two different animals have a similar form, the same function, and similar location, everyone refers to them by the same name, and no one doubts their fundamental identity. These are *analogous organs*. But further observations soon showed that in some cases, even though the analogy is obvious, the form and function may vary widely. Among vertebrates, for example, the anterior member can be a leg for walking, a wing for flying, or a fin for swimming. Its form has changed and its function modified, but the parts still retain the same general form, and when these parts have undergone certain modifications, the position of the limb and its relationship to the other organs have remained essentially the same. Geoffroy assumed that what is true about some anterior members is equally true for all the other organs as well. In 1806, this hypothesis led him to identify the external structure of the fins of fish with the legs of other vertebrates, and this permitted him to relate the composition of the skulls of all these animals to a common type. On seeing how valuable this guide could be, he finally put forward his *principle of connections*. 'An organ,' he said, 'may become altered or atrophied, but it is never moved to another location.'[3] This was based on an entirely new concept of organs, and much of our subsequent thinking about anatomy was based on this principle.

---

[3] *Philosophie anatomique* Introduction: xxx, 1818.

Until then, it was generally thought that an organ is designed to perform a particular function, but Geoffroy said that, on the contrary, the organ exists independently of the specific function it serves in a given animal. For him, the concept of a structural design and what we would call today *morphology*, is more important than the notion of *physiology*. The animal has a basic structure that is always the same, regardless of the role it may play in nature. The functions and forms of an organ are determined by the animal's faculties and the conditions under which it must use them. This way of viewing living creatures was a definite step forward.

From that time on, there was a more fruitful approach to anatomy that Geoffroy Saint-Hilaire did not hesitate to use in his studies of embryos. When one compares the head of boney fish with that of adult mammals, one sees very quickly that the former has a great number of bones that have no analog in the latter. This seemed to present a serious obstacle to the theory of a unified plan of composition. Geoffroy had the brilliant idea of comparing the heads of fish not only with those of adult mammals but also with those of the mammals' embryos. In this way he could identify not only the bones in these animals but also the centers of ossification and how they are related to one another. This enabled him to make the appropriate comparisons and show that, despite their apparent differences, there are unquestionable resemblances between the bones in the heads of fish, reptiles and birds, and those of mammals. Pursuing this approach, Geoffroy discovered rudimentary teeth in the jaws of very young whales and in the embryos of birds that, as adults, lack teeth. How delighted he would have been had he been able to foresee that paleontologists would one day exhume true birds whose teeth were not only as well developed in the adult state as they are in mammals but seemed to be lost as the birds matured!

The fish with its multiple head bones and the bird with its teeth that disappear almost immediately and become part of the surrounding tissues, could be thought of as having been arrested in their evolution toward a stage of final development that has only recently been reached by mammals. These considerations led Geoffroy to interpret them as though they were permanent embryos of the higher animals. Bonnet, Erasmus Darwin, and Diderot had foreseen a kind of parallelism between the embryonic development of animals and the successive modifications of the species. Geoffroy's comparison of the lower animals and the embryos of higher animals provided a precise way to test the parallelism that Serres and Henri Milne-Edwards were seeking. And it is exactly the same idea that was expressed by Fritz Müller and the partisans of embryonic development when they said: 'The successive forms that an animal presents in its embryonic development are only a condensed repetition of the forms through which its species has passed before arriving at its present state.' This formula is certainly too simplistic: the embryonic forms of an animal could hardly live outside the egg. They are usually modified by the presence of a more or less voluminous nutrient vitellus, by various adaptations, and especially by the accessory phenomena that determine how quickly development takes place by what we have called *embryonic acceleration*. Nevertheless, Fritz Müller's law is still one of the fundamental principles of comparative embryology, and it is basically nothing more than a generalization of what Geoffroy Saint-Hilaire had already pointed out.

But if the development of lower animals often tends to follow that of the embryos of higher animals of the same group, then whenever the latter are, for some reason or other, retarded in the development of some of their parts, those parts should exhibit characteristics that are appropriate for the lower forms of their family.

In 1820, Geoffroy made this idea the basis for a new science, teratology, which undertook to classify the abnormal, often hideous forms of animals and relate them to the standard laws of embryogenesis. Throughout time these strange *monstrosities* have been subjects of countless bizarre legends. Instead of considering them simply as exceptions to the normal laws of nature, Geoffroy used them to discover, elaborate, and verify those laws. He demonstrated that monstrosities can always be attributed to some identifiable physical cause and even went so far as to show how one could create certain types of monsters experimentally. Camille Dareste has recently pursued experimental studies of monstrosities.

Most of what we call monstrosities are simply the result of an arrested development of certain parts of the animal, but they can also come from a welding of organs that normally remain separated. Studies of the latter led Geoffroy to an important principle that is as valid and useful in comparative anatomy as it is in teratology: 'Welding can take place only between similar parts.' It seemed to Geoffroy that these parts have a mutual attraction for one another, and he was so impressed by this that he wanted to make it a general rule. Toward the end of his life he considered it one of the fundamental principles governing the combination of matter. He thought that this mutual attraction governed all aspects of closely related bodies, just as the universe is governed by the mutual attraction of astronomical objects.

Unfortunately, if the facts that he took as basic precepts were valid, the causes to which he tried to attach them were nothing more than an illusion. Organs that share fundamental characteristics do not necessarily have a particular attraction for one another. If they frequently weld themselves together, this is due to their being born symmetrically on each side of the body; they dispose themselves where it is easiest for them to develop. If, for any reason, their growth is more rapid than that of the parts that separate them, they are commonly found in contact with one another. From then on, their tissues become melded just as the stem of a plant does when grafted to the trunk of another plant.

If monstrosities are attributed to natural causes and are simply the result of important modifications during the normal progress of development, is it not possible that these modifications could be common and manifest themselves not only in individuals born of the same parents but also in their descendants? If the laws of teratological development are only special cases of more general laws, is it not possible that as soon as these monstrosities appear, they could perpetuate themselves, multiply, and take their place among the forms that reproduce themselves and become a new zoological species? This idea that *abrupt variations* could come about in such a manner must have occurred to Geoffroy Saint-Hilaire, and in the course of developing his ideas, he must have realized that birds could have been separated in this way from reptiles:[4] 'A reptile, in the earliest stages of development, may have experienced a contraction toward the center of the body in a way that separated the blood vessels in the thorax and base of the pulmonary sack in the abdomen from the rest of the body. This could have provided the appropriate conditions for development of a bird.' Today, it seems unlikely that sudden modifications of this kind could have been a common cause of diversification, but in the case of birds, paleontology has fully confirmed their genetic relationship to reptiles as was proposed almost simultaneously by both Lamarck and Geoffroy Saint-Hilaire.

Up to this point, Geoffroy directed all his efforts toward studies of vertebrates, and fish, reptiles, birds, and mammals became the principal focus of his research. For this branch of the animal kingdom, which was generally considered the most important one, the unified plan of composition became a firmly established principle, and in his pursuit of this goal, Geoffroy continued to make new observations and unexpected discoveries. Anatomists now possessed a systematic method of investigation by which discoveries are no longer the result of chance but products of systematic research. Rigorous precepts were established for comparative studies of organs, and morphology was freed from the restricted role it had played in physiology. Embryology was introduced as a fruitful source of information for all aspects of the anatomical sciences. Studies of the structure of higher animals were now subject to precise laws, and these laws governed even the small deviations that Geoffroy had thought of as caprices of nature. Work of this kind could not be limited to one part of the animal kingdom, as important as it might seem; it had to extend to the entire realm of animals.

In 1820, Geoffroy Saint-Hilaire undertook a study of the articulated animals. He may have been inspired by the remarkable work in which Savigny, his friend and companion during the expedition to Egypt, had described these animals in a memoir that was already a classic. Savigny had shown that the seemingly highly varied mouths of the coleoptera, flea, bees, flies, and butterflies always consist of the same parts that are similarly placed even in the most diverse species. In the coleoptera it is designed for grinding, in bees for licking, in fleas and flies for piercing, and in butterflies for taking in liquids. In a series of important studies published in 1820, Audouin applied the method of analogs to all parts of the bodies of arthropods and thought that he could show that the same number of parts are found in equal numbers in all parts of the body. He said that 'all the differences that one sees in the segmented animals are only a function of the similar or dissimilar growth of the segments, the union or separation of constituent parts, or of their degree of development, which is complete in some and rudimentary in others.[5] Latreille had just shown that all the appendages of the arthropods are nothing

[4] *Memoires de l'Académie des sciences,* vol. XII.

[5] *Annales des sciences naturelle*, vol. I, p. 116.

other than modified legs and even included in that definition the wings of insects, thus relating them to the respiratory appendages of crustaceans and aquatic arthropods. The uniform plan of composition of the articulated animals, that is to say the arthropods, gained a foothold in science at the same time that a uniform plan for the composition of vertebrates was accepted. The moment had come to try to show that these two groups were really only one.

There are profound differences in the structure of the nervous systems of the vertebrates and the arthropods. In the former, it is entirely dorsal; in the latter, it is mainly ventral except in its anterior part where it is crossed by the digestive tube and forms a kind of ring, the *æsophageal ring*. Inferences based on this æsophageal ring might seem, at first sight, to indicate that the nervous systems of the vertebrates and the articulates have opposite connections and that it would be quite impossible to relate them to the same plan. Geoffroy wondered[6] whether the solution to this problem might not lie in the marked contrast of the configurations of these nervous systems. How are the regions that we call the *back* and the *belly* of an animal defined? The belly is a region of the body that faces the ground, whereas the back faces the sky. To identify these two regions we should not view them in relation to the animal itself, as the principle of connections would require, but to the external surroundings. It may be that what appear to be different arrangements of the organs of the arthropods and vertebrates may only be a difference in the attitude that the two animals take in their posture. If one places a vertebrate with its back down and stomach up, the apparent difference between the vertebrate and an arthropod immediately disappears. One sees that the organs actually occupy the same relative positions and that the anatomies of the vertebrate and arthropod are in fact very similar. In both cases, the three main organic apparatuses, the nervous, digestive, and circulation systems, are found to occupy exactly the same positions with respect to one another. Even within a given group, the posture of the animals is far from constant. Geoffroy cites a certain number of fish, insects, and crustaceans that have an attitude that is the direct opposite of that of other members of their genetic group. (We shall later have occasion to make several additions to this list.) Thus there is nothing inconsistent with well-established facts in assuming that the attitude of the vertebrates is the reverse of that of the arthropods. Embryology has since added further evidence in support of Geoffroy's interpretation.

The illustrious anatomist was on less solid ground when he wanted to pursue more detailed comparisons in the hope that he might discover the significance of certain parts of the skeletons of articulates or find equivalent parts in vertebrates. As early as 1692, Willis had concluded that the muscles of arthropods are enclosed by bones. Hoping to find solid parts of insects that would be analogous to those that seem characteristic of the vertebrates, Geoffroy noted that, among the arthropods, the solid arch of the carapace that protects the body repeats itself just as regularly as the vertebrae in the skeletons of higher animals. He did not hesitate to

---

[6] *Ibid.*, 1820, p. 462 and 539.

interpret these parts as truly analogous. After that, an extraordinary conclusion became inevitable: the bodies of vertebrates enclose their spinal column, but the opposite is true of articulates. How can such a strange disposition be explained?

Geoffroy began by pointing out that, contrary to what one might think, articulates are not unique in this respect. In the case of the turtles, certain parts that seem to be analogous to parts of the internal skeleton of other vertebrates are closely welded to the carapace, so that these animals are also partly enclosed in their skeleton and in this sense can be considered transitional to the articulates. But Geoffroy sensed that this simple comparison would not be convincing unless he could find an explanation for it. He thought that the development of all organic systems has been influenced by the relationships of the circulatory and nervous systems. Among the vertebrates, these two systems work together in harmony and contribute to the development of all the organisms that are able in this way to reach their highest degree of perfection. In the molluscs, the blood system predominates, and the animals remain soft as if they were saturated by liquids. In insects, the circulatory apparatus is rudimentary, and it is the nervous system that controls the course of development. The parts that are most closely related to this system – and the skeleton is one of them – are the first to develop and become complete long before the others become well established. Those that form in the vicinity of the nervous system and develop more slowly will necessarily be enveloped by it to produce the articulate. It is obviously not necessary to go into a detailed discussion of this *à priori* explanation that is very similar to that of Oken and the *natural philosophers*. It is based on the simple hypothesis that the nervous system and circulatory apparatus intervene in the processes of development.

However that may be, once Geoffroy had come to consider the solid cutaneous segments of the arthropods equivalent to the bodies of the vertebrates, he could see nothing but the sides in the members of these animals. The arthropods walk on their sides which, instead of forming a continuous circle as in most vertebrates, would be openly exposed. According to Geoffroy, these sides would have analogs only in pleuronectid fish, and, from the point of view of their skeletons, the crustaceans and the insects should be considered to be walking on their flank, while from the point of view of the nervous system they walk on their back. It has always seemed rather difficult to reconcile these two points of view that Geoffroy was able to entertain simultaneously and remain convinced of the value of his method. He also pointed out other homologies between the arthropods and lower vertebrates: the heads of insects are formed by three segments, like the cranium of the vertebrates; their wings, which, according to Latreille, are modified respiratory organs, correspond to the swimming-bladder of fish; the marks are still preserved on the latter in the form of the small regularly placed orifices that follow the lateral line. Impressed by these apparent resemblances, he exclaimed:

'Yes, there can be no doubt; I can now affirm that these beings which, until now, were thought to be without vertebrae will have to be included among the vertebrate animals in our natural series.'

This conclusion seemed attractive to a number of eminent thinkers: Oken and Goethe in Germany were prepared to accept it, and in France, Latreille also wanted to compare the crustaceans and fish. On the 10th of January 1820, he read before the Academy of Sciences a memoir in which he tried to show that a crab, if considered only externally, is a kind of fish in which the opercular or jugular region has not grown in the manner of the thorax while the rest of the body is divided into segments. In 1824, Ampère, the illustrious physicist to whom we owe electro-magnetism, expressed himself in an anonymous letter published in the *Annals des sciences naturelles* in which he modified and improved on Geoffroy's seminal idea. He saw in the skeleton tegumentary components equivalent to the sides of the vertebrates, and, in his view, the rachidian canal of arthropods has remained open on the upper side; the spinal marrow has disappeared, and the ventral chain, which fills its functions, corresponds to the system of the sympathetic ganglion of the vertebrates. Thus, all contradictions and seeming anomalies in the comparison between the vertebrates and articulates disappeared, and the convergence of the two types seemed a more reasonable way to make the idea more readily acceptable. One could cite a long list of illustrious persons who, while expressing certain reservations, agreed with his basic idea.

When an idea arouses this much interest and leaves such a profound impression in the minds of scientists that it survives, despite certain factual evidence that seems contradictory, it is generally true that the idea expresses a truth that is still imperfectly perceived. Between the vertebrates and the arthropods, there are two points of indisputable resemblance: the vertebrae of the former are exactly like the rings of the latter. The principal organs of both have the same relative disposition if, instead of considering their orientation relative to the ground, one considers only their orientation with respect to another part, such as the nervous system.

Those are the facts as we know them. It is now a matter of explaining them and, if we can, interpreting their meaning. Geoffroy and his contemporaries always considered vertebrates to be the typical animals and tried to find all their parts in the lower animals. This was where they went wrong. There is no more reason to look in the lower animals for all that is found in the higher ones than there is to look in the egg, or even in the embryo, for all the organs observed in the adult animal. But if we now know this, it is in part because we have used the methods of comparison that Geoffroy introduced to science when he was careful to relate the lower animals to the embryos of the higher animals, and because he contributed more than anyone else to overturning the doctrine of genetic assimilation [emboitement] of the germ cells that was still upheld by Cuvier. It is also because, along with Lamarck, he valiantly defended the idea of the mutability of specie without which evolution would be impossible and without which the idea of gradations in the organic complexities would remain confused and sterile. Thanks especially to the discoveries of Semper and Balfour, we can now consider it an established fact that the bodies of primitive vertebrates were segmented much as they are in the articulates and that all that was required for segmented animals to become vertebrates was a reversal of their primitive posture: the reason for this rotation has become quite clear,[7] but one can be sure that there is no essential resemblance between the exoskeleton of the arthropods and the internal skeleton of the vertebrates. Moreover, it is not only the segmented animals that have a well-developed external skeleton, and it is not the arthropods that the vertebrates resemble. As the weak development of the skeleton in the lampreys and amphibians might lead us to expect, it is with the soft articulated animals and annelid worms that they appear to have the closest affinities.

Deeply impressed by the marked resemblances shared by the higher animals and having seen in his studies of monstrosities that external conditions could influence the final outcome of evolution, it is only natural that Geoffroy became convinced of the mutability of specific forms. At a time when Cuvier's work had led scientists to recognize how many forms have become extinct, Geoffroy, the person who more than anyone else was responsible for the development of anatomical philosophy, began to share Lamarck's impressions and wondered whether it might not be necessary to accept these ancient inhabitants of the Earth as the probable ancestors of present animals. From 1825 to 1829 he published several memoirs on the fossils of giant reptiles found in the vicinity of Caen and Honfleur. He showed that these animals, to which he gave the names *Teleosaurus* and *Steneosaurus*, are quite distinct from modern crocodiles. But if one accepted this first point, it raised another question, namely: "if the so-called crocodiles of Caen and Honfleur are in the same terrestrial Jurassic beds as the *Plesiosaurus*, could they not provide an unbroken temporal and anatomical link that attaches these very ancient inhabitants to presently living reptiles known as gavials.[8] Without being categorical, Geoffroy did not hesitate to accept the possibility of a similar transformation, for, he said, 'ecological conditions have a great power to alter organized bodies.'[9] and he added a few lines later: 'In my opinion, respiration constitutes such a powerful control for the disposition of the animal forms that there is no need for the respiratory fluids to be abruptly and extensively modified, in order to bring about scarcely perceptible alterations of the forms. The slow action of time is ordinarily enough to accomplish this, and the effect is even greater if the creatures have survived a catastrophic event. From one century to another, the growing number of subtle modifications produce a cumulative effect. If respiration becomes increasingly difficult to the degree that the organs can no longer deal with it, the animal spontaneously creates another arrangement, and perfects or alters the pulmonary system by modifications that may be either *favorable* or *fatal.* These modifications lead to changes of other organs as well and influence the animal's entire organization. *If these modifications lead to harmful effects, the animals that experience them will cease to exist and will be replaced by others with forms that are only slightly changed but changed in better accord with the new conditions.'*

---

[7] On this matter of the vertebrates and segmented animals, see my work *Les colonies animals et la formation des organisms* p. 662 to 700.

[8] *Recherches sur les Sauriens fossiles*, p. 4.

[9] *Influence du monde ambiant sur les formes animales*, p. 76.

This statement is of special importance, because it clearly defines the difference between Lamarck's doctrine and that of Geoffroy Saint-Hilaire. Lamarck saw the external world acting on living creatures only through the intermediary of what he took to be their habitual behavior, and the entire organism becomes involved in the animal's modifications. Geoffroy, while not rejecting Lamarck's ideas completely,[10] considered the organism to be passive and saw in the successive modifications of living forms the direct effect of their environment. For Lamarck, as for Buffon, the major cause of the extinction of living forms was man. These two naturalists did not think it likely that species would disappear unless their behavior had an adverse effect on humans. Geoffroy, however, thought that species can disappear naturally when their organization is no longer in accord with the environment in which they must live or when they have undergone unfavorable modifications. The statement cited in italics above shows that he attributed this disappearance to a true natural selection that is a direct effect of the environment. It is not provoked or stimulated by a rapid increase in the number of individuals and the greater struggle for survival that this entails. The spontaneous disappearance of species without cataclysms is a well-established fact and is clearly a factor in another important phenomenon, the formation of new species.

This appearance of new species may come about in more than one way. In addition to the gradual, imperceptible modifications mentioned in the passage cited above, Geoffroy believed that there are also abrupt modifications such as those he considered responsible for the transformation from reptiles to birds. Modifications of this kind resemble those that, even in ordinary times, can produce monstrosities. In other words, a monstrosity with exceptional characteristics that are fortuitously related to a new mode of existence in a particular environment can take root and become the origin of a new species or even of a distinctly new type derived without transitional forms from a distinctly different source. Geoffroy wondered why nature has not used some of the phenomena we commonly see reproduced so clearly in the course of embryonic development as a means of producing a greater diversification of species.

This similarity between the embryonic development of the individual and the evolutionary development of specific types is considered, for good reason, one of the most brilliant discoveries of zoology, and it is a relationship that Geoffroy always had in mind. Note how he described and interpreted the metamorphosis of amphibians: 'We witness every year,' he says,[11] 'a spectacle that is perceived not only by the perception of the mind but by that of the body as well, a spectacle in which we see organized amphibians transform themselves and pass from the organic conditions of one class of animal life to that of that of another. A creature

---

[10] Geoffroy was especially critical of the choice of *special proofs* on which Lamarck had based his doctrine. As for the influence of habitual behavior on organic modifications, no physiologist would question its importance. It would be easy to cite a great number of organic forms that have been fixed by heredity in an attitude that is most habitual for them and has led in turn to important organic modifications.

[11] Memoire sur l'influence du monde ambient pour modifier les formes animals, 1831, p. 82.

starts as a fish in the form of a tadpole then becomes an amphibian in the form of a frog. We now know how this marvelous metamorphism comes about. It demonstrates the hypothesis presented above that the transformation of an organic form is to the next, immediately higher stage.

'The physiological aspects of this transformation of the tadpole have been collected and accurately described by my famous friend M. Edwards,[12] in his work entitled: *De l'influence des agents physiques sur la vie*, and the anatomical aspects have been described by many naturalists, most notably by Dr. Martin Saint-Ange.

'This transformation results from the combined action of light and oxygen, and the changes of the body are due to the production of new blood vessels that are essential to achieving a harmony between the organs. If the fluids of the circulatory system are directed into new channels, less remains for the older ones. These alternative vessels, which contract in some places while expanding in others, change the relationships of the organs wherever they occur, and, as the transformation proceeds through all parts of the body, it becomes more pervasive, in some places by atrophying and losing some parts and by hypertrophy of other parts where there had been scarcely any primordium to build on. When Dr. Edwards kept the tadpoles under water, he retarded or even prevented their metamorphosis. What was done on a small scale in this experiment nature was done on a much grander scale in the case of the proteus that inhabits the underground lakes of the Carolines. Deprived of light and the ability to breath the open air, it remained in the perpetual state of a larva or tadpole. But actually, it could easily have transmitted to its descendants the ability to adapt to these conditions which may be the very ones that governed the existence of the first reptiles when the world was totally submerged.'

Geoffroy not only recognized the influence of the environment, but like Bacon before him, he recommended experimental research to determine what conditions led to permanent modifications. He cited experimental evidence, such as that of modifications of domestic animals that had been transported to America, that had not been fully appreciated and exploited. 'Naturalists of our time,' he said,[13] 'are so impressed by the isolated descriptions of morphology and natural phenomena and so accustomed to taking their dissecting scalpel into the internal labyrinth of the organized beings, that they seem to fear that they may compromise themselves by studying the relationships and reciprocal actions of parts of the universe. This research is difficult enough by itself, largely because of its originality, but it is eminently meaningful and likely to contribute progress.'

This is the program to which Charles Darwin has so magnificently contributed. Geoffroy clearly recognized the reciprocal influence that beings living side by side must have on one another. He also foresaw that the effects of modifying an organ could not be isolated from the rest of the body. He thought that there are some

---

[12] This refers to William Edwards, brother of Henri Milne-Edwards, and presently the doyen of the Faculty of Sciences of Paris.

[13] *De l'influence des circumstances exterieures sur les corps organisées*, p. 26.

organs that develop in unison and others that are reduced by the growth of another. Many correlations of this kind could be found, and there are even more in which the concomitant modifications have been dominated by the modification of a single organ. Thus there is room for more studies *'of the organs that act together with an exceptional degree of sympathy and have a strong influence while others play only a secondary role as official associates.'* It is clear that Geoffroy had the notion of these *coordinated modifications*, to which Charles Darwin regretted having failed to attach sufficient importance in his last publications. In 1835, in his *Etudes progressives d'un naturaliste*,[14] Geoffroy expressed his views about living creatures and their origin when he said: 'In my view, there is only a single system of creations that are incessantly reworked, perfected, and integrated with previous changes under the all-powerful influence of the external world.'

During the same period, another brilliant naturalist, Cuvier, upheld and ably defended opinions that were exactly opposed to those of Darwin and Geoffroy Saint-Hilaire. This led to a heated debate, the story of which deserves to be told, because it was not without profit for natural philosophy and casts light on the value of doctrines that might otherwise have long remained sterile.

[14] See note on the first page of the Preface.

Chapter X

Georges Cuvier

Affinities with Linnaeus; the influence of Cuvier on scientific work; global revolutions; theory of successive creations and migrations. – Cuvier's inferences. – The order of appearance of animals; special creation of the principal groups. – Natural classification: adherence to the principle of ultimate causes; the principle of environmental conditions; law of the correlation of forms; law of subordination of characteristics. – The four major branches of the animal kingdom.

We have just seen that the views of the two great naturalists, Lamarck and Geoffroy, shared a close intellectual relationship to those of Buffon. They considered almost all the philosophical aspects of the natural history of animals and developed marvelous insights based on their ability to synthesize from an intimate knowledge of zoology. With admirable logic, they developed insights drawn from other branches of science that enabled them to explore a wide range of ideas and make them consistent with a single, supreme goal: to discover the secret plan of creation. In a similar way, Cuvier expanded on the work of Linnaeus.

The background of this person who would one day dominate the natural sciences through his brilliant discoveries and intellectual innovations was quite different from that of Geoffroy. While still a student in Paris, Geoffroy undertook to pursue his studies of the higher vertebrates under Daubenton, whereas the young George Cuvier, who at that time was a tutor for the Héricy family at the chateau of Fiquainville near Fécamp, was devoting his leisure time to studies of the lower animals, mainly the invertebrates that flourish in the sea. There was no attractive central plan that Cuvier could build on. The class of worms, in which Linnaeus had included almost all marine invertebrates except the crustaceans, was a highly diverse assemblage of species that did not seem to have anything in common other than their low place in the ranks of animal life. Starting in 1795 when he was only twenty-six, Cuvier proposed that this chaotic class be abandoned and its invertebrate members - animals with white blood, as Aristotle called them – be distributed into six classes, namely the *molluscs, insects, crustaceans, worms, echinoderms*, and *bryozoa* [zoophytes]. This proposed classification recognized the important resemblances and differences within a group of animals that until then were very poorly known. It is remarkable that the classification of invertebrates that Cuvier proposed when he was just starting is closer to what we accept today than the one

A. McBirney, S. Cook and G. Retallack, *The Philosophy of Zoology Before Darwin: A translated and annotated version of the French text by Edmond Perrier*,

that he arrived at many years later. The impressions one has as a young man tend to be stronger and are often closest to the truth.

Cuvier was impressed by the marked differences he saw between white- and red-blooded animals and was convinced that they are separated by a profound gap. He never retreated from this view. His mind was closed to the idea of a unified animal kingdom that Geoffroy adopted and made a guiding principle for the rest of his life.

Cuvier's first memoir, published in 1795, contained hints of correlations that he, like Aristotle before him, foresaw and would later go on to establish more firmly. At the time, however, his works, like Aristotle's, scarcely mentioned these relationships. For example, all the white-blooded animals that have a heart are distinguished by also having limbs; those that have no heart but only a dorsal vessel breathe by means of trachea. All those that possess a heart and limbs also have a liver; the others do not. Cuvier used these correlations for purposes of classification without attempting to explain them. He simply stated them as laws of nature based on observational evidence, and this cautious approach became increasingly characteristic of his work as he advanced in his career as a naturalist.

These early results, which were communicated to Geoffroy Saint-Hilaire in 1794 while Cuvier was still living in Normandy, aroused the young professor's enthusiasm. He wrote to his future rival: 'Come join us. You can take the role of a new Linnaeus.' It was indeed another Linnaeus that had appeared, but a very wise Linnaeus who would employ the strict laws governing the distribution and organization of animals to rejuvenate a field of study that for centuries had been fading. It would awaken the imagination of Cuvier's contemporaries and enable them to contemplate in a new light the tangled debris of past studies.

Pursuing his research on the lower animals, Cuvier produced a series of memoirs on the anatomy of limpets (1792), the anatomy of snails (1795), the structure and types of molluscs (1795), a new type of mollusc – the nudibranch [phyllidies] (1796), the anatomy of the lingulas [a kind of brachiopod], the anatomy of the *sea squirts* [ascidies] (1797), the blood vessels of the *leeches* [sangsues] (1798), the red-blooded worms (1802), the *sea slug* [aplysie], the *sea feathers* [vérétille] and corals (1803), the *salps* [biphores] (1804), and various pteropod molluscs [with wing-like flaps] or sea slugs. At the same time, he explored the history of vertebrates, and assembled valuable documents on the bones of antediluvians that were beginning to turn up in many places.

Finally, in 1811, Cuvier compiled all his studies of extinct animals in a major work modestly entitled: *Recherches sur les ossements fossiles*. The preface to this work has become famous under the name *Discours sur les revolutions du globe*. It was here that he laid out the general conclusions he had drawn from his studies of the origins and ancient history of the animal kingdom. Written in a very clear, elegant, and grandiose style, it could not fail to make a great impression. In the coming years it would have a strong influence the direction of geological and paleontological research and in some cases even dictated the conclusions based on this new work. Cuvier was determined to let the facts he accumulated speak for themselves and made it a practice to state only their most immediate consequences. He categorically rejected all theories and encouraged us to use sound reasoning to deduce the past in terms of all possible systems. And most important, he introduced rigorous methods of investigation into natural history. As one reads his scientific masterpiece, each step of his reasoning seems absolutely sound and the progression toward a final conclusion seems beyond question. His method of assembling facts and coordinating them in terms of a general concept, soon became the ruling feature of an important group of scientists. It was formally presented as a scientific method, and it is interesting to consider the effects it had when Cuvier introduced it at the beginning of the nineteenth century.

The large ruptures of the crust Cuvier saw in great mountain chains, along with the folding, faulting, and profound stratigraphic unconformities, convinced him that terrible cataclysms have repeatedly disrupted the surface of the Earth. Who would not have had a similar impression on contemplating, for example, the tormented rocks of the Pyrenees with their overturned and distorted beds and the deep chasms that look as if a giant sword had cut into their flanks? The evidence is very impressive. It seems that nature has not yet had time to repair the disorder left by its latest convulsions. Cuvier was so impressed with the image of terrible cataclysms that he became obsessed with the inevitable effects that he believed these events must have had on life on Earth.

The impact of these events on animal life seemed to be beyond doubt. The proof was evident in the bodies of rhinoceroses and mammoths preserved in the ice of Siberia. These animals must have been frozen instantly after being killed. Otherwise, decay would have consumed their flesh and left nothing but skeletons. But where do rhinoceros and elephants live today? In the torrid climate of Africa. The climate of Siberia must have been just as warm at the time these animals lived, and the same events that caused them to perish also made the climate of the land they inhabited glacial.

'This catastrophic event,' Cuvier added, 'occurred instantly with no gradational beginning, and it could scarcely have been less dramatic than all the earlier ones that proceeded it. The rupture and upheaval of older beds leaves no doubt that sudden, violent movements were responsible for the conditions in which we now find them. Evidence of these forces on great bodies of water are still visible in the masses of fragmental detritus and coarse conglomerates that are inter-bedded with many of the strata. These terrible events must have occurred repeatedly. Living beings without number were victims of these catastrophes: some that lived on dry land were engulfed by deluges; others that lived in water were stranded on land when the seafloor suddenly rose. Entire races were wiped out, and the only traces they left for the naturalist to study were bits of almost unrecognizable debris.

'The marks that these events left in almost every part of the world enable us to interpret the story of what must have happened.' Cuvier states his belief without reservations: how could the factual evidence be more compelling and his interpretation be more reasonable?

After laying out this idea that sudden, violent events have led to global revolutions, Cuvier tried to demonstrate that they could not be explained by normal, on-going processes. He quickly reviewed the effects of the rain, wind, and running water, the waves of the oceans, volcanic phenomena, and earthquakes and dismissed their importance. He hesitated to rule out the possible influence of the modifications of the position of the Earth's axis only to say: 'The magnitude of such phenomena is disproportionate to the scale of the effects we have just noted. In every case, their very slow action rules out any possibility that they could explain catastrophes of the kind that must have occurred.' These forces are inadequate to explain the present state of the Earth's crust, and the explanations for the inferred revolutions are buried in a mystery that will be very difficult to unravel. As for the duration of the period of tranquility during which human history has unfolded, Cuvier carefully reviewed the archeological evidence and estimated it to have been of the order of six thousand years.

We know where geologists stand on this today. All agree that the present period has lasted about five hundred centuries[1], and all recognize that almost all the present structure and topography of the earth's crust result from phenomena very similar to those that operate today. Everything confirms that these phenomena have been slow and gradual, that there have never been general cataclysms nor sudden revolutions. It has been clearly shown that the elephants and rhinoceroses buried in the ice of Siberia were adapted to live in cold countries.

Everything we now know directly contradicts Cuvier's conclusions. How can one explain that, at a time when Geoffroy Saint-Hilaire and Lamarck already recognized the processes responsible for the present state of the Earth, Cuvier's eminently logical and precise mind remained closed to them? A dominant theme of his *Discours sur les revolutions du globe* is his conviction that science is sometimes confronted with long-standing enigmas and that it is useless to look for their meaning. Cuvier took pleasure in showing the weaknesses of the explanations proposed by the great names of the past. Descartes, Leibnitz, Kepler, and Buffon, along with Robinet and Telliamed, were all targets of his criticism. Their basic ideas and the factual observations that are in accord with them are completely ignored. But the human mind has an irresistible need to synthesize facts and deduce their meaning. This intrinsic need that man has always felt became the essential element of reasoning that makes him the man he is. When the mind is presented with two facts simultaneously, it instinctively assumes they have a cause and effect relationship, even though it may be quite unintelligible. If there is no theoretical principle to warn us that there could be other facts worth considering, the omnipotence of a divine will is taken as the explanation for the relationship, and the explanations no longer seem unreasonable. The mind accepts all the consequences that seem to be implied by the union of two elementary facts, no matter how absurd they may seem.

---

[1] See, for example, Credner, *Traité de géologie*, French translation, p. 255.

If Cuvier had been less impressed by the weakness of our understanding of nature and less convinced that the absurd systems of Leibnitz and Buffon held nothing of tangible value, and if he had had less disdain for the general concepts of his time, he might have been less inclined to believe that a region of the earth could have been suddenly plunged from a torrid climate into a glacial regime. He might have asked himself whether elephants and rhinoceroses similar to those found in Siberia had been adapted to live in the warm countries where the species are now confined. His attention would have been drawn to their thick coats of hair, and he might have discovered that, as we now know, mammoths lived among herds of reindeer and at the time of their death were already well adapted to cold climates. Siberia was not suddenly covered by ice but had been that way for centuries. He may have had doubts about the sudden nature of the cataclysms he had deduced and may even have thought them improbable. Had the ideas of Lamarck and Geoffroy about the slow changes that take place on the surface of the earth been better known, we might have been spared a line of reasoning that became a burden to many branches of natural history.

Today, no one believes that these great cataclysms and sudden global changes ever took place. In developing ideas of this kind, it is better to proceed cautiously by weighing the most direct consequences of rigorous observations and assessing them in terms of firmly established knowledge. For example, if one fails to consider the time interval represented by a break in the stratigraphic record, one might conclude that different faunas were suddenly modified and succeeded one another abruptly in certain stratigraphic suites. On seeing the uniformity of faunas during the primary period one immediately concludes that the climate was the same throughout the world and that the seas were the same everywhere without wondering whether this uniformity might not mean that there had not yet been sufficient time for more varied species to develop in different ecological conditions. Would not the world's present fauna and flora appear to have a monotonous uniformity if we were to omit the dicotyledon and monocotyledon plants from our present flora and all the mammals, birds, reptiles, amphibians, bony fish and insects from the fauna? Would not the climates seem to have suddenly changed? Left without the thermometer that these creatures provide, we would be unable to judge climatic differences. Who knows whether the statements about uniform global temperatures during that early period merit any more confidence than would those dictated by the hypothetical circumstances I have just outlined? Many more examples of this kind could be cited to illustrate the dangers that exaggerated claims pose for science. Instead of allowing a soaring imagination to dominate our judgment of these questions, we should keep our wings folded and search for a more secure route through the labyrinth of facts.

What would be the effect of such periodic cataclysms on plants and animals? Cuvier thought that each revolution caused a great number of species to disappear. This was in contrast to Lamarck who considered man to be the only animal capable of destroying the work of nature. How are the species that disappear from the Earth replaced? Is it necessary to have a totally new creature? This is what Cuvier proposed, and many have subscribed to his belief. He does not state this explicitly in the *Discours sur les révolutions du globe.* In fact, he seems to have consciously avoided it when he wrote: 'when I say that the strata contain the bones of several species that no longer exist, I do not claim that a new creation was necessary to produce the species that exist today. I simply say that they did not exist where we see them now and that they must have come from elsewhere.'

But this passage applies mainly to man and to the higher animals, notably the mammals, for Cuvier stated elsewhere that the various classes of animals have appeared successively, and this assumes that each has been the product of special creation. After laying out the order in which the fossils occur, he says that 'it is reasonable to assume that fish and shellfish did not exist at the time when the primordial world was formed, and it seems that the oviparous quadrupeds must have appeared along with fish when the earliest secondary beds were formed. The terrestrial quadrupeds did not appear, at least in considerable numbers, until much later when the thick limestone beds had been deposited…'

After the deposition of these thick limestones, one finds only 'detrital beds, such as sandstone, marls, and shales, that indicate a more or less tumultuous transport instead of the earlier, more tranquil deposition. There are irregular beds of coarse, rocky fragments just below or above the fluvial beds, and they generally show signs of having been deposited in fresh water.

'Almost all known cases of viviparous quadrupeds of this kind are either in these fresh-water deposits or in these transported beds, and as a consequence there is every reason to believe that the quadrupeds did not come into existence, or at least did not leave their remains in the beds that we have been able to examine, until the next to last retreat of the sea and under the conditions that preceded its last flooding.'

Cuvier thought – or to use his way of putting it – was inclined to think that each of the great zoological groups that we have just enumerated was the object of a special creation. As for species, he thought that they had remained unchanged since their creation. He could consider this a demonstrable fact, because he believed he had established that the present period had lasted only 6000 years and that animals preserved since Egyptian time differed in no way from present animals. The argument, however, obviously loses much of its strength if the duration of the present period must be increased at least ten-fold as geologists believe would be more realistic. In addition, even with regard to the fixity of species, Cuvier had reservations. If the higher animals are really fixed, it could well be that the animals with white blood are as well. Wishing to explain why his paleontological studies were focused mainly on mammals, he wrote: 'Shellfish clearly show that the sea existed where they were formed, but their changes of species could very well result from minor changes in the nature of the water or just in temperature.' One can, of course, interpret this passage as referring to migrations of species rather than modifications of their morphology, and what follows seems to make the former more likely. But at the beginning of his discourse, Cuvier was most explicit when he expressed himself as follows:

'It is reasonable that life could not have remained unchanged when the water in which it lived was changing… The animal world must therefore have undergone a succession of variations that were occasioned by changes of the water in which it lived, and even if change of the water were not responsible for the variations, they at least corresponded to them in time. It is these variations that have gradually brought the aquatic animals to their present state.'

This passage obviously tells us something about what was going through Cuvier's mind, but when a writer of his ability leaves us with an equivocal sentence of this kind, we can take it to mean that his opinions were not yet firm, and that is all we should try to make of it.

We find similar signs of uncertainty in his considerations of the species from which his *Animal Kingdom* originally developed:[2]

'There is no proof that all the differences that distinguish the organized beings today could have been produced by ecological changes of this kind. All that has been said on this subject is hypothetical. Observational evidence seems to indicate that, on the contrary, *in the present state of the world*, the variations are confined within rather narrow limits and as far as we see back in time, those limits have been the same as they are today.'

If he had stopped there, Cuvier could have remained consistent with the facts, but he went on to generalize and arrived at a conclusion that does not necessarily follow from the limited amount of observational evidence:

'One is forced to conclude that certain forms have perpetuated themselves *since the beginning* without exceeding fixed limits, and all beings that share one of these forms constitute a *species*. Varieties are just accidental variations of the species.

'Since the observed variations are the only way we have of knowing the limits to which varieties can extend, one must define species as the grouping of individuals that have descended from common parents and resemble those parents as much as they resemble one another.'

In summary, Cuvier firmly believed that sudden upheavals had affected the surface of the globe and that these upheavals destroyed the greater part of the species living in the affected region. Later, these species were replaced by others that could migrate from regions that had been spared. It is not necessary to create new species after each cataclysm, but it is not impossible, and in any case there can be no doubt that the different classes of animal life that have appeared were created in an orderly succession. Marine species have been able to survive most of the events that disturbed the emergent continents, but the composition of water has no doubt undergone numerous changes over time, and this has brought corresponding modifications of the assemblage of species living in a given locality. This was the essence of Cuvier's theory. As usually happens, it was exaggerated by some of his disciples, several of whom have turned the hypothesis of *successive creations* into an inflexible dogma that calls on special creations at each major geological period.

---

[2] Le règne animal distribué d'après son organization, 1829 ed., p. 9.

It matters little whether plants and animals have been created once and for all or whether the productivity of the creative power has been manifested repeatedly. From the moment that one accepts, as Cuvier did, the belief that species are fixed and immutable and that each one must have been the product of a distinct creative act, one need no longer be concerned about their origins. All Cuvier's efforts were turned in another direction, namely the ways in which many animals have unquestionable similarities while others have profound differences. Cuvier forced himself to organize these differences in a very orderly manner. He tried to relate the resemblances according to the same laws that govern the organization of the animals' anatomy. In devising a natural system for the classification of animals, he was also laying a foundation for comparative anatomy.

One of the most pressing needs of the Linnaean period was a method by which the supposedly immutable species could be clearly distinguished from one another. The main goal was to find a simple way to recognize species that had been described so that names could be given to those that had not. One could not assign names to new species without recognizing their differing degrees of resemblance and seeing that the species of plants and animals form long series in which the characteristics of successive members differ only in minor ways. Extreme forms that at first seem so different from one another are found to be linked by great numbers of intermediaries. It was this gradational relationship that was the basis of Bonnet's scale of living forms, the uniform plan of composition favored by Buffon and Geoffroy Saint-Hilaire, and Lamarck's concept of descent through evolution. It is also what led Linnaeus, Jussieu, and Cuvier to conceive the idea that nature has a plan of creation and that that plan must be the basis for our system of classification. In listing species, each one must be placed between the two species that resemble it the most. Once one knows the place of a plant or animal in such a classification it should be possible to deduce the organization of the group to which it belongs. This system, which was referred to as a *natural method* is clearly distinguished from the artificial systems with which, for want of anything better, the first classifiers had to be content.

The search for this natural method, which Linnaeus considered one of the greatest problems that had to be resolved, has been the principal concern of many naturalists. Jussieu strived to establish the principles that would make it possible to apply this method to the classification of plants. Cuvier, convinced that the method was scientifically sound, defined these principles and applied them with exceptional clarity to the animal kingdom. 'In order for a method to be good,' he said, 'it is necessary that each being have its own inherent character. One cannot draw on characteristics that are transitory patterns of behavior; they must be drawn from shared traits.' These simple words completely eliminate the embryology to which one looks today for the solutions to all the difficult problems of affinity, and what, in the near future, will reasonably be a major revelation of the true genetic relations of animals. Anatomy becomes the exclusive basis of classification.

But which of all the different features that can characterize the organization of an animal should one employ in setting up the major divisions? As Cuvier pointed out, all characteristics do not necessarily have the same value. He said that 'it is the sharing of such traits that serves to excludes others; it actually necessitates it. When one identifies certain traits in a being, one can determine which of them are shared by a group and which are incompatible with it. The parts, properties, or shared traits that have the most incompatible relations with others or exercise the most marked influence on the lives of beings as a whole, are what are called *important dominant characteristics.* The others are *subordinate characteristics* and there are different degrees of them.'

Naturally, it is the most influential characteristics that will define the most extensive divisions, and others will follow in order of importance. In short, that amounts to saying that there are sets of characteristics for major branches, classes, orders, genera or species, an idea that was evidently in Linnaeus's mind when he set up his hierarchy of zoological and botanical divisions. But aside from this principle of *subordination of characteristics* that was the essential element of his method, the passage I have just cited expresses another principle that Cuvier used as a basis of comparative anatomy. It is the *principle of correlation of parts* which expresses a dual concept: first, that the parts of a living creature are so closely linked to one another 'that none of them can change without the others changing as well,'[3] and, second, that, given the form of one of an animal's organs, one can deduce from it the forms of all the others. There are some very bold propositions in this that may not be as narrowly restricted as Cuvier's words might lead one to suppose. To use Cuvier's example, if one considers the body of an animal to be a function of several variables, that function will appear *à priori* to be so complex and the variables so numerous that one will be led to conclude that there will ordinarily be multiple solutions and that many of them will be indeterminate. Cuvier got around this problem by calling on another principle that seemed capable of resolving it, the principle of *conditions of existence* which holds that each animal possesses all the faculties it needs and only those that are essential to its survival in the conditions in which it must develop. This proposition naturally leads to the principle of the correlation of parts and is nothing but the *principle of final causes,* a principle that Cuvier considered particularly relevant to the natural sciences and, in his view, the only valid basis for scientific reasoning.

When he came to apply these principles, Cuvier found that he was obliged to descend from the heights where his excessive ardor had led him, and in the end he said of the principle of correlations of forms: 'As most would agree, this principle is rather self evident and needs no further demonstration, but when it is a matter of applying it, there are many instances in which our theoretical knowledge of the relationships of forms is insufficient and requires more observational evidence.... Since these relations are constant, they must have an appropriate cause, but since we do not know what that is, we must look for further observations to make up for

---

[3] *Discours sur les revolutions de la surface du globe* Didot Edit, p. 62.

the inadequate theory. This enables us to establish empirical rules, and when these rules are based on large amounts of observational evidence, they are almost as reliable as those based of theoretical reasoning.' This is a good illustration of the difference between the methods used by Geoffroy Saint-Hilaire and Cuvier and helps us appreciate the different consequences that they entailed. Geoffroy considered various possible explanations for the relations he saw between various parts of an organism, but Cuvier never allowed himself to indulge in speculations of this kind. Even if he did not understand all aspects of these causes, Geoffroy tried to establish as much factual evidence as possible, so that he could extrapolate from what was known and predict very remote relationships in the organs of creatures that live today. Cuvier, on the other hand, declined to be guided by this kind of incomplete knowledge and was obliged to follow the observed facts methodically as far as he could without venturing beyond them. Not only did he willingly deny himself use of a valuable procedure, his exclusive reliance on existing facts exposed him to faulty interpretations arising from uncertainties in the paleontological and geological relations that he was unaware of.

Geoffroy suspected that the embryos of whales and birds might contain the primordia of teeth and on searching found them. He foresaw the discovery of birds that had teeth, such as the *Hesperornis* or *Ichthyornis* that were found in American chalk beds. Cuvier, however, could not have anticipated such a discovery as long as he remained faithful to his method. If someone had given him the jaw of a bird with teeth, the principle of correlation of parts would have prevented him from relating that jaw to anything other than a reptile. Geoffroy, like all men who are inspired by a general unifying idea, whatever it might be, was in the privileged position of an observer who stood on a high promontory where he could view a vast panorama. The landscape with all its towns, forests, fields, mountains and valleys was laid out before him so that he could see all their detailed relationships to one another. Cuvier could rise to a higher viewpoint when he thought it necessary but recommended against scaling great heights. In his view, one must always keep one's eyes on the closest objects, proceed slowly, and never attempt to proceed without first exploring every pathway on foot. When he addressed Geoffroy, he sounded like a lion advising the eagle never to use its wings.

In reality, the principle of correlation of parts has always remained in the realm of metaphysics. The paleontological methods that Cuvier had used to make his discoveries consisted of only a careful comparison of fragments of fossil skeletons that he had at his disposition with the corresponding skeletons of living animals. This comparison required a sound knowledge of science that only a person with Cuvier's ability could bring to bear on the problem. As we have seen many times, this method has many dangers, but Geoffroy's theory of analogs had provided a precise method of investigation that has since been adopted by all anatomists.

In zoology, Cuvier followed a strict plan based on the principle of subordination of characteristics. When he tried to identify 'the must influential characteristics that should be the basis for the primary divisions,' he proceeded with an *à priori* assumption. 'It is clear,' he said, 'that the most important features are those that govern the animal's functions, namely, its sensations and movements, for they are essential to the creature's existence and define its animal nature.'[4]

Cuvier addressed himself first of all to the nervous system. He attached special importance to this system and went so far as to say: 'The nervous system is basically the essence of the animal; the other systems are there only to respond to and serve its needs.'[5] He recognized that the nervous system is present in four different states in the animal kingdom: (1) it may be an assemblage made up of the brain and the spinal marrow both of which are enclosed in a bony envelope; (2) it may consist of material dispersed through the vital organs and joined by nervous filaments; (3) it may be formed from two long ventral cords joined by a collar of two ganglia situated in the esophagus; or finally, (4) in the case of certain animals, the nervous system may to be very indistinct. On the basis of his observations, Cuvier summed up his ideas about the animal kingdom in the following passage.

'If one considers the animal kingdom in terms of the principles just laid out and, disregarding long-standing prejudices, takes into account only the intrinsic nature of the divisions independently of any accessory factors, such as their size, usefulness, or how much we know about them, one finds that there are only four basic plans. This way of classifying animals seems to reflect the way they have been organized in nature, and all the other divisions that naturalists have proposed are nothing more than minor modifications based on the development or addition of features that do not alter the basic plan in any essential way.'

When we do this, we find that it reduces the scope of the unified plan of composition. There would be four distinct plans rather than a single one with no intermediate transitions between them. But why four? Why not more or fewer? Cuvier does not seem to have been troubled by this question. For him, observations were all that mattered, and the facts are self-sufficient; they call for no discussion, explanation, or interpretation. There are four ways in which the nervous system is organized, and they correspond to four main branches of animal life. That is all the reasoning one needs. But how can one overlook the implications of this reasoning? What it really means is that *the nervous system is the basis of the entire animal and that the other organs are there only to maintain it and serve it.* This proposition, to which no anatomist or embryologist could subscribe today, was regarded by Cuvier as a self-evident axiom; but that is because he had deduced it himself, not from observational evidence, but from other principles that were essentially metaphysical.

If species were immutable and created independently, it would be natural to assume that there is a system that regulates the organs and governs the development of the constituent, unchangeable parts of each individual. This system is a faithful agent of the Creator's will; it is the nervous system. The will of the Creator is present in all the 'germs' and though it remains indivisible, it maintains the relative size and position of this system within the whole body during its growth.

---

[4] Le Règne animal 2nd edit. 1829, vol. I, p. 46.

[5] Annales du Muséum d'histoire naturelle, 1812, vol. XIX, p. 76.

These parts already exist in the germ cell as simple miniature versions of the individual from which it is detached, and only those parts that have remained hidden need to grow and develop in order for the new-born individual to become identical to its parent.

Thus, in Cuvier's system, everything gravitates around this idea that, apart from sudden cataclysmic revolutions that he believed he had demonstrated, all nature is immutable. Extinct species that were closely related to those living today had the same habits and lived in the same climates as they always have. The present species have always been what we see today. The individuals themselves, despite their apparent changes and metamorphoses, only develop parts that have long been latent, and they all have a germ containing a reduced image of the organism from which they are detached. The nervous system is a repository of the fundamental form of each type and regulates the growth and order of appearance of the parts that, on evolving, cannot diverge from a path they have followed through all eternity. The various organic types are arranged according to four different dispositions of the nervous system. It is hardly surprising, since his species could not change, that there were no transitional forms and that these four types are completely isolated from one another.

How different these ideas are from those of Geoffroy! As the author of *Philosophie anatomique* saw it, our world has undergone a slow evolution, and the types of changes that have occurred are no different from those that are effective today. Species modify themselves in response to changing climatic and ecological conditions but only little by little. During its lifetime, the individual never ceases transforming itself. In the egg, its parts develop gradually one upon the other just as each branch of a tree is produced from the one that supports it. The circumstances in which this development takes place can influence its development and allow for the appearance of new forms that are linked in the same way that the members of a succession of animals follow one another.

For Cuvier, each living creature is the miraculous work of a will that designed and produced it. For Geoffroy it is the result of a long series of phenomena closely linked to one another. It was inevitable that two doctrines as opposed as these were would lead to conflict. In 1830, a solemn debate brought them face to face in the Académie des Sciences.

Chapter XI

Debate between Cuvier and Geoffroy Saint-Hilaire

Attempts to extend the theory of a uniform plan of composition to molluscs. – Cuvier's opposition; what is the meaning of such a uniform plan? – The relationships between embryology and epigenesis are clarified. – Cuvier's insistence on the pre-existence of the germ cells. – Von Baer and the four types of development. – The school of ideas versus the school of facts. – The influence of Geoffroy Saint-Hilaire, Cuvier, and Lamarck.

On the 15th of February, 1830, Geoffroy Saint-Hilaire read before the Académie des Sciences of Paris (under Latreille's name as well as his own) a report on the works of two young naturalists, Laurencet and Meyranx, who had set out to demonstrate that the physiological organization of cephalopod molluscs[1] could be related to that of the vertebrates. In 1823, Latreille had taken a special interest in this subject and had pointed out several types of external resemblances between squid and fish. De Blainville had also attempted to make comparisons of this kind. Laurencet and Meyranx delved more deeply into the question and tried to find connections between the various organs of a cephalopod similar to those seen in vertebrates. To do that, they had to resort to an ingenious fiction. They assumed a vertebrate that was doubled over at the level of its navel, so that the face of the ventral side remained outside and the two halves of the back were welded together. The pelvis was close to the neck, and the limbs were attached to an extremity of the body, so that when the animal was walking on these limbs it took 'exactly the same position as an acrobat who turns his head back over his shoulders and hands.' The intestines of the cephalopods are doubled over, pieces of cartilage in the rear of their neck are connected to what is called their funnel, and eight or ten limbs around the head serve as arms and legs. These features are so characteristic that they offer a natural explanation and an unexpected way to place the most elevated of the molluscs at the level of vertebrates. The parrot-like beak of the cuttle-fish and its large, complex eyes serve to make these analogies even more convincing. As extraordinary as the explanation of Laurencet and Meyranx may appear, it did not come as a surprise to many naturalists; several scientists, including even some of those who were most closely attached to the Cuvier school, had often had

[1] These are the octopus, cuttle-fish, calamary, and their analogs.

recourse to more drastic means than simple folding to contrive an analogy between beings that are only remotely related. The embryonic development of animals is also rich in strange phenomena of this kind.

The Academy might have adopted the report of its commission without discussion if Geoffroy Saint-Hilaire had not insisted that the work of Laurencet and Meyranx lent support to his own ideas. He cited a passage in which Cuvier, after having enumerated all the characteristics that distinguish the cephalopods from fish, arrived at the conclusion: 'In a word, we see that, regardless of what Bonnet and his followers may have said, nature has jumped from one plan to another leaving a distinct hiatus between two of its products. The cephalopods are not transitional to anything: they have not developed from other animals, and they have not developed into anything higher.' Cuvier saw the conclusion of his colleagues' report as an attack on his own published work.

Over the years, the two illustrious naturalists had stated their opposition to each other's ideas more and more pointedly and from many different points of view. More than once, Cuvier had used his reports on the work of the Academy to criticize rather bitterly the views of his former friend, and as early as 1820, Geoffroy ended his memoir on articulated animals with these moving words expressing the grief that Cuvier's words had caused him:

'Some may well think that I should refrain from reporting these facts in order not to offend persons in their mature years whose long experience has left them less receptive to seductive ideas. The younger people whom I am addressing here are naturally hungry for novelty. Scientific integrity, a love of truth, and a resolve not to dissimulate compel me to warn these young men about the consequences of results of the kind I report. I can offer them no higher sign of respect than to warn them that when they become passionately committed to views they take to be of great philosophical interest, they should be aware that these same views have been rejected (with some violence) by the leader of the modern school, the greatest naturalist of our time [i. e. Cuvier].'

The time had come for these two adversaries to stop their skirmishing and face each other in a formal confrontation. This time Cuvier responded to Geoffroy Saint-Hilaire's report with a frontal attack on his unified plan of composition, which he attempted to show does not exist.

'In any scientific discussion, the first thing one must do,' he said, 'is to clearly define the terms one employs... Let us begin by agreeing on our understanding of these important words: *unity of composition* and *unified plan*.

'The *composition* of something means, in ordinary language, the parts of which that thing is composed, and the *plan* means the arrangement that these parts have with respect to one another.

'To take a simple example that will illustrate the idea more clearly, the *composition of a house* is the number of rooms it has and its *plan* is the layout of those rooms. If each of two houses has a vestibule, an entranceway, a bedroom, a salon, etc. on the same floor and arranged in the same way, so that one passes from one to another in the same order, one would say that *their plan is the same*.

'... But what is the *unified plan*, and especially, the *unity of composition*, that we are told should henceforth serve as a new basis for animal life? These words obviously cannot be employed as they ordinarily are in the sense of *identity*, for a polyp or even a whale or snake does not possess all the same organs arranged in the same way as they are in humans. As they are normally used, the words 'unity of plan' or 'unity of composition' mean nothing more than *resemblance* or *analogy*. But once these extraordinary terms have been properly defined and stripped of their nebulous meaning, we see that they are being used in a distorted sense, and instead of furnishing a new conception of zoology that all could agree on, they replace one of the most basic principles on which zoology has rested since it was first established by Aristotle.'

Thus, for Cuvier, not only was there no such thing as a uniform plan, but he maintained that Geoffroy's doctrine and methods were nothing new; they went back to Aristotle, the father of philosophy. Of these two accusations, one is well-founded, the other obviously unjust. There is no question that the unified plan of composition for the entire animal kingdom could not be sustained in the same sense that its defender presented it. Geoffroy may have been a bit hasty in presenting a concept he had painfully derived from theoretical reasoning. The only thing that could be said was that the author of the *Philosophie anatomique* may have perceived a much more extensive relationship between animals that were closely related in appearance than any one before him had been able to see. These resemblances were not confined to a small number of shared characteristics; they were also found in the details of parts that grew, shrank, or merged with other parts, as well as in their various transformations. It involved comparing animals not only in their adult state but at all stages of life, and in order to do this Geoffroy devised a method, the *method of analogies*, the rules of which had never before been formulated. The basic approach of this method, as others have pointed out, is independent of his uniform plan of composition. Whether there was a single plan of organization or several, the rule applied to all animals constructed on the same plan and became such a valuable guide that Cuvier's successors have continued to make it the normal tool for their research. It enabled one to recognize the prevalence of plans of organization in nature and included not only the general principle governing the relationships of species but also comparisons with embryos. Because Cuvier, a partisan of the pre-existence of germ cells, could not appreciate the full importance of embryology, Geoffroy was able to extend the notion of a plan of organization farther than Cuvier, and he could do this without departing from the rigorous definition demanded by his adversaries.

Geoffroy clarified and added support for the principle of connections by using another principle, possibly of more general importance, on which he based certain aspects of his comparative embryology: *all the organs of an animal are formed from one another in a specific, constant order*. It followed from this that the organs of adult animals will always have to present the same relationships.

But, as we have seen, Geoffroy believed that this development resulted from the dual influence of the nervous and circulatory systems, the effects of which cannot be the same in all parts of the organism. The external conditions in which this development takes place may intervene and alter the effects, and they may cause some of the organs to remain in a primitive state, while others atrophy and even disappear. Some organs do not develop at all and allow their neighbors to grow to a relatively exaggerated size. This results in displacements, merging, or disassociation of various organs, in what appear to be deviations or even a complete departure from the common plan. But the plan can always be found if one rigorously applies the principle of connection not only to comparisons of adult animals but also to their embryos at differing stages of development. In other words, it is necessary, as Geoffroy clearly proposed, to search for unity not so much in the definitive result of the animal's development as in the manner in which this development is accomplished. This enabled Geoffroy to get around much of Cuvier's argument and justified applying his theory not only to very simple creatures but also to others with a very complex organization. The development of simple organisms has remained largely incomplete. He says quite clearly:[2] 'The molluscs have been placed too high in the zoological scale. If they are considered to be only at a very early stage of development involving fewer of their organs, it does not then follow that they lack the organs that would be expected to appear in the development of successive generations. Organ A will have an unusual relationship with respect to organ C if organ B has not developed, and if the development of the latter is arrested prematurely, this relationship may not appear at all. This is how the organs can develop different spatial relationships and lead to a variety of observed configurations.'

This simple statement shows the importance that comparative embryology, a new field of science to which Cuvier made only cursory allusions, must have had in Geoffroy's zoological research. And the results were everything the founder of anatomical philosophy had hoped for and even more. In fact, the explanations for the phenomena explored by this new research are still based on Geoffroy's original precept of a general plan on which all animals are constructed. Varieties result only when the development of a number of parts is either prematurely arrested or proceeds at an excessive rate. In truth, the unity of plan, as Geoffroy envisaged it in the case of vertebrates, is only a *result*, and when he made it one of the goals of nature, Geoffroy did just what he accused Cuvier of doing – mistaking the effect for the cause. Nevertheless, a fruitful new approach had been opened. Observation would soon become the only sound source of factual evidence, and it is through Geoffroy's theoretical work that we have come to recognize the necessity – or at least the importance – of a new kind of observations.

At one stage when Von Baer was making observations of this kind in Russia they seemed to lend support to Cuvier's views. Von Baer believed that he could recognize four types of development in animals, corresponding exactly to those that anatomy had indicated to Cuvier. But one of the *à priori* arguments that Cuvier invoked against the uniform plan of composition can also be turned against

---

[2] *Principes de philosophie zoologique* 1830, p. 70.

his own system: 'If one goes back to the origins of all living things,' he said,[3] 'what other law could be a better constraint than the necessity of according to each being the means of ensuring its existence? Why would the Creator not have been able to use a variety of materials and instruments to do this?' This is true, of course, but why would the author of all these marvelous things have gone on to create four distinct plans rather than a single one? This is what modern science is beginning to perceive. I have tried to show in my work on colonial animals that there have been geometrical constraints on these relationships, but to do that I had to modify Cuvier's concept in significant ways. Just as Geoffroy had deduced the principle of a uniform plan of composition mainly from his studies of only the vertebrates, Cuvier had been led to conceive of the existence of four main branches through his study of only relatively advanced animals. Von Baer had proceeded in essentially the same way; the four types, when stripped of their lower forms, must then have seemed to him extremely neat and absolutely distinct. It was not long, however, before numerous aberrant forms began to turn up. Some of these could be traced back to an ideal progenitor, but others have resisted any connection to a prototype, and it was obviously necessary to recognize that the characteristics of the four branches could be lost in the lower forms. There are real transitions between certain branches of animals, and the only thing the animals of one of these large divisions had in common was a similar disposition of parts of the body that were otherwise dissimilar. Although each series could be related to a simple form, the early forms were found to lack the distinguishing features of that particular group. These early, more primitive forms were found to be much less distinctive. This was the conclusion of work that, as we shall see, was carried out over the following years.

Although Cuvier was getting closer to reality than Geoffroy Saint-Hilaire, his belief in the existence of four distinct types of organisms was no longer strictly correct.

By now, the differences between the two academicians were proving to be more profound, and the questions they raised were becoming increasingly fundamental. As one authoritative scientist wrote:[4] 'From the day in 1806, when Geoffroy Saint-Hilaire undertook to illustrate the unity of composition by a method based on a *combination of observation and reason*, he passed from *analysis* to *synthesis*, and it was at this point that he planted the seed for all his future scientific disagreements with Cuvier. But neither he nor anyone else foresaw the kind of plant that seed would produce. The two colleagues still believed that they shared a common view, but an inevitable dissension was already taking form. One of them was becoming an innovator, while the other had to become either a disciple or an adversary. Cuvier was not the sort of person who could be anyone's disciple, and

[3] Article on NATURE from the *Dictionaire des sciences naturelles*.

[4] *Vie, travaux et doctrines scientifique d'Etienne Geoffroy Saint-Hilaire* by Isadore Geoffroy Saint-Hilaire, p. 375.

because their ways of thinking were so different, this was especially true in the case of Geoffroy Saint-Hilaire. He could only become his adversary.'

Cuvier had not always refused to synthesize. His *Discours sur les revolutions du globe,* which he wrote as the introductory chapter of his *Règne animal,* is an obvious example. But little by little, his growing disagreement with Geoffroy led him to formulate in a reasonably clear manner his increasing opposition to his colleague's ideas. As he said in 1829,[5] 'I have had a long-standing commitment to an objective examination of established facts.' Later, he recommended that any naturalist worthy of the name should restrict himself to the facts and should never venture beyond the immediate consequences of those observed facts. The sole preoccupation of the true naturalist should be to name, classify, and describe. For him, this was the only way to guard against errors. He declined to pursue his discussions of Geoffroy's doctrine at the Academy of Sciences but continued to expound his views at the Collège de France where he presented a series of brilliant lectures on the history of the natural sciences and the successions of schemes that, over time, had impassioned the human mind and had briefly led us down blind alleys before ending up in the waste-bin of science.

When offered by a man of such distinction, lessons of this kind, can have a resounding effect, but if we restrict science to the mere collection of facts we are in danger of throwing out many humble but worthy contributions. When important new concepts are turned aside by barriers erected by those seeking to discredit them, genius is put at the mercy of anyone who can use a magnifying glass or scalpel. By denying the freedom to reason one risks wasting all the efforts to explore mysteries and question dogma for the mere sake of flattering personal vanities. This would sacrifice what is most personal in man, his right to create new ideas. Such intentions were certainly far from Cuvier's mind, but acts of this kind have their inevitable consequences. Would this great man, who had brought us such magnificent concepts, have wanted to see his name used as a banner for the *school of facts* that was viewed with increasing disdain by Geoffroy's increasingly enthusiastic disciples? Geoffroy could not remain indifferent to this assault. He gathered all his energy against the claims of the self-styled 'positive' school – a term soon be coined – that they were maintaining natural history in accordance with 'traditional customs of the past'.

'Certain types of minds,' he said,[6] 'can be convinced only by what they see with their own eyes, not by deduction drawn from the consequences of such observations… They take this stance in order to avoid innovative thinking; they will consider nothing but physical perceptions that they can deal with without losing touch with things that are palpable to our senses. For this school, the natural sciences have three missions: to *name, record,* and *describe.*

---

[5] *Mémoire sur l'Hectocotyle.* By a bizarre coincidence, in this same memoir which was supposed to contain nothing but proven facts, Cuvier arrived at the erroneous conclusion that the hectocotyli, which is now known to be a simple arm of the octopus, is a kind of parasite.

[6] *Mémoire sur l'oreille osseuse des crocodiles et des téléosaures*, 1813, p. 136.

'This school, which has become particularly prevalent at this time, teaches us that the history of science testifies to many examples of theories that have descended one after the other into the immense abyss of human error, that ideas are nothing in themselves, and that only facts can withstand changing interpretations and survive. However, instead of giving the youth of humanity over to the derisive criticism of our present society that is concerned with nothing but the influences of the moment and the advance of civilization, would it not be better to explain these natural vicissitudes in a way that would allow one to see them in an historical context? And as for this claim that the facts they presented constitute the whole domain of science, I believe it would be more correct to say that the facts will be able to reach future ages only if they are guided and supported by ideas that are relevant to them and give them their principal value.

'Facts, even when carefully recorded by an intelligent observer, can never be of much value to science if they have been presented in a piecemeal and haphazard manner. Since we should throw as much light as we can on this thesis, let me illustrate it with the following parable:

'Suppose that Paul had the will and means to procure all the pleasures of life: he was intelligent, inventive, and he applied himself to studies and to assembling all that he thought might be of use in achieving this end. He stocked his cellar with the very best wines, he filled his woodbin with all the fuel he needed to warm his house, and he took steps to ensure that all his needs could be met. He always chose the best quality and stored everything neatly and securely, so that orderliness prevailed throughout. But that was as far as Paul would go. He never drank the wine he had stored, and he never warmed his house with the firewood. He left all these fine things nicely arranged and never used them. – You will say to me *Paul was a madman*, and I shall agree.'

But Paul was not completely mad. It sometimes seemed that he would never be able to accumulate all the possessions he needed to fulfill his dream. A time came when, without foreseeing it, he could no longer enjoy life. Having always made himself out to be very wise, he continued to see wisdom in this endless accumulation and could not prevent himself from treating as foolhardy those who, like him, had assembled the materials they needed but realized that the moment had come to put them to use.

The open struggle between the two colleagues was brought to a sudden end on the 13th of May 1832 when Cuvier suddenly died. Geoffroy then had to defend himself from all those who believed that it was their duty to pursue their master's mission. He must have regretted that he no longer had his illustrious adversary to deal with, and it surely saddened him to have to read the petty arguments of Cuvier's disciples that only deepened the grief he felt for his lost friend. His inner suffering is revealed in the following characteristic passage:

'I can no longer pursue these discussions that I formerly conducted under more pleasant circumstances. I have now become the toy of those who oppose me and use the somber loss I have suffered to discredit me in my last days… It is painful for me to leave these imperfect pages that I would like to have brought to a proper conclusion, but the harassment that I have had, combined with the ravages of age, leave me discouraged and unable to change my behavior during these last hours of my life. My better judgment and frailness tell me to turn away from the struggles that some would like to engage me in.'[7]

Three years earlier, when he was still filled with courage and enthusiasm, Geoffroy had written: 'There is more to science than collecting facts... we must exercise judgment in order to understand those facts. It will then be said, as it often is by those around me, that such judgments amount to theorizing. I prefer not to let myself be bothered by this argument that is more noise than logic, and my response to all this drivel, which seems to have no purpose but to drown out the opposition, is that the time for spouting poetry and making vague accusations has passed. These cries speak for themselves. They are nothing but *declamations*.'[8]

Matters did not sort themselves out as quickly as Geoffroy thought they would. Even today, many scientists still wonder whether naturalists are justified in using syntheses of the kind that are widely used with such facility by physicists and chemists. Many, especially those whose early studies were focused on man, still consider the animal kingdom inexplicable. They oppose all attempts at coordination, and even go so far as to say that they are impossible. In 1821, Geoffroy had given them this severe warning: during a discussion of the chances that the armies of the republic could force a passage of the Rhine, a veteran officer of the ancien régime had convinced his listeners that such an enterprise would be folly. He had scarcely ceased speaking when the news arrived: the French troops had just accomplished the impossible – they had crossed the Rhine.

Cuvier, despite all that he said, did not put all his trust in facts, and, similarly, Geoffroy was always careful to avoid becoming entangled in theoretical aberrations, such as those of the German school that we shall examine shortly. If he tried to deduce the secrets of nature, he did so methodically, and his 'premonitions' were almost always examined in relation to factual observations that exercised a control similar to that of experimental studies. This approach and his methods of study are now an integral part of the modern *anatomy* and *experimental zoology*. He may be accused of engaging in dangerous theoretical speculations, but they do not diminish in any way the value of his method or the importance of his synthesis. The close alliance of observation and reasoning exercised a strict control on his thinking. It has been expressed very well in the words of the illustrious German scientist, Johannes Müller:[9]

---

[7] *Notions de philosophie naturelle*, 1837, p. 111. Geoffroy had just been removed from the management of the Museum's menagerie he had founded and replaced by Frédéric Cuvier.

[8] *Etudes progressives d'un naturaliste,* 1835, p. 84.

[9] Johannes Müller *Handbuch der Physiologie des Menschen*, II Band, p. 522: Die wichtigsten Wahrheiten in den Naturwissenschaften sind weder allein durch Zergliederung de Begriffe der Philosophie, noch allein durch blosses Erfahren gefunden worden, sondern durch eine denkende Erfahrung welche das Wesentliche von dem Zunfälligen unterscheidet, und dadurch Grundsätze

'The most important truths of the natural sciences are not found through a simple analysis of a philosophical concept nor by a single observation; they are the result of well-planned experiments that separate the essential from the accidental and reveal fundamental laws from which one can then deduce numerous consequences. It is not just experimentation; it is philosophical experimentation.'

This is also the opinion of Henri Milne-Edwards.[10]

'In some schools, theoretical speculations are viewed with great disdain, and one is constantly told that only facts have importance in science. But to me that seems to be a grave error. Such thoughts would be excusable if they were from an obscure workman who is employed to construct a large edifice from natural materials and believes that the role of the architect consists only of piling the stones one atop the other. He would see the plans prepared with an artist's pencil as nothing more than a trick of his imagination, a useless fantasy. But the workman in the quarry, if he follows the products of his labor and sees that all the shapeless blocks that he has produced have been assembled under the hand of a master to construct the Parthenon of Athens or the Coliseum of Rome, would understand that the architect's plans are not useless, even when the monument that resulted from his genius may have only an ephemeral lifetime and the debris of the building that has now fallen in ruin serves only as a source of material for new constructions.'

Science, no matter how one approaches it, cannot accommodate itself exclusively to the methods advocated by only one of two different schools. Those who claim to adhere to the facts are always pleased to get new ideas and hasten to put them to good use. On the other hand, one rarely sees the authors of a theory present it as anything more than a guide in the search for new discoveries that afford a better understanding of the relationships of established facts. Everyone now agrees that the best method to follow is to imagine the possible result of experiments or observations before undertaking them; to design the study so that it will enable one to chose between the possible hypotheses inspired by known facts in order to see which one best conforms to reality. These hypotheses are then used to acquire new information that will enable one to arrive more or less directly at an explanation of the workings of nature. Unfortunately, man is not just a reasoning animal; the accord that could easily be reached if he confined himself to the exercise of such reasoning is often disturbed when he allows his passions to come into play. In fact, the supposed discord over methods that still arises from time to time only serves to conceal ambitious vanities or petty personal quarrels.

From that time on, the natural sciences were able to follow a more productive path. Thanks to Cuvier, a new form of science was created that studies plants and animals of the past in order to provide a detailed history of the Earth. Even though Cuvier seems to have deliberately declined to explore the consequences of this new approach, the doctrines of Lamarck and Geoffroy served to open his horizons.

---

findet, aus welchen viele Erfahrungen abgeleitet werden. Dies ist mehr als blosses Erfahren, und wenn Man will, eine philosophische Erfahrung.'

[10] *Leçons de physiologie et d'anatomie comparée*, 1857, vol. 1, p. 2.

It was simply a matter of using a rigorous examination of the factual evidence, together with the deductions one could logically draw from it, to determine the origin of all that has inhabited the Earth. The hypothesis of a uniform plan of composition led Geoffroy to create his theory of analogs and gave comparative embryology the importance and clear direction it lacked until that time. At the same time, however, Cuvier's opposition prevented the broader implications of Geoffroy's seductive hypothesis from being generally accepted. It also drew attention to several other types of organisms and called for more serious studies, mainly of lower animals, that, as we shall soon see, rejuvenated the field of zoological philosophy. When Lamarck introduced the idea of a gradually increasing complexity of different types of organisms and proposed a possible relationship between these types, he revealed the potential importance of heredity. And when Cuvier insisted that these same groups to which Lamarck had attributed so much mutability were in fact immutable, he aroused great curiosity about them and demonstrated the need to find an explanation for the long stability of certain species and their isolation in nature.

Thus, Milne-Edwards' beautiful image of three different edifices constructed by each of these men of genius was clearly in need of revision, but elements of their separate contributions would eventually be incorporated into the definitive theory that would soon emerge.

Chapter XII

Goethe

Goethe's thoughts on the unity of types of organization. – Metamorphosis and structure of plants: the ideal plant. – Studies of comparative anatomy; research on the ideal type of skeleton. – Goethe's conception of descent with modification [transformisme] – Kielmeyer.

An important yet simple idea, such as that of a unified plan of composition, was like a breath of poetry that spread throughout all science. More than one partisan of Geoffroy's doctrine must have seen this unity as a revelation of a omnipresent divinity that is guiding continual change throughout the universe and enjoys astonishing us with its infinite variety of combinations, all of which display overwhelming evidence of their origin.

Cuvier complained to Geoffroy that 'a confused kind of pantheism is hidden behind your theory of analogs.' This is precisely why the theory, which was condemned in France, gained an ardent defender in Germany, the illustrious Goethe.

While taking up the cause under Geoffroy's banner, Goethe still maintained his staunch originality. Even when he was younger and before Geoffroy had begun his brilliant scientific career, Goethe had deduced a new, robust concept and skillfully developed it. Struck by the modifications that cultivation can produce in the various parts of a plant, the botanist La Hire and most notably Linnaeus had proposed more or less explicitly that these modified plants were basically the same as the original ones from which they had developed and under the right conditions could be transformed from one to the other. This interpretation is expressed in the following passage of Linnaeus' *Philosophie botanique*. 'Flowers, leaves and shoots all have the same origin… The perianth is formed by the union of rudimentary leaves. A luxuriant vegetation inhibits the growth of flowers and transforms them into leaves, whereas a sparser vegetation modifies the leaves and transforms them into flowers.'[1]

The same idea is found in these sentences extracted from his *Aménités academiques*: 'A shrub that produced flowers and fruit each year when growing in an earthenware pot will cease to bear fruit when transplanted into fertile soil and will develop nothing but branches with leaves. The branches that formerly

[1] Linné, *Philosophie botanique*, Edit. Gleditsch, p. 361.

bore flowers will now be covered with leaves, and the leaves in turn will give way to flowers if the plant is replaced in a pot where it finds less nourishment.'[2]

Several naturalists, Ferber, Dahlberg, Ulmark, and especially Gaspard Wolf expanded on the observations of the Swedish naturalist, even though they did not concur with all his conclusions. They warned that some of Linnaeus's interpretations conceal a number of traps under a seductive outward appearance.

Goethe grasped this basic idea and expressed it with the same clarity that is found in all his work. In 1790 he showed, not, as is often said, that all parts of the flower and other parts of the plant are only transformed leaves, but that the petals, stamen, and various parts of the fruit are only transformations of an organ the primitive nature of which he was seeking to determine. 'It is clear,' he said, 'that we shall need a single general term to designate the fundamental organ that undergoes these metamorphoses and to which one can relate all the secondary forms.' But Goethe never proposed such a term, and his theory was absorbed into science in a more restricted form that took the leaves to be the organ from which all the others are derived. Goethe enlarged on his theory in the following proposals:[3]

'We know that there is a strong analogy between a shoot and a seed, and we also know how easy it is to discover in the shoot the outline of the future plant. Although it is not as easy to recognize the presence of roots, they exist no less than the seeds and develop just as easily and promptly under the influence of moisture.

'The shoot has no need of cotyledon, because it is completely attached to the mother plant. As long as it is attached or has been grafted on to another plant, it draws its nourishment from it, but when it is placed in soil, it quickly develops roots.

'The shoot is composed of a series of buds and leaves in various stages of development from an earlier stage of evolution. *The branches that come from buds on the stem can be thought of as young plants still attached to the parent plant, just as if they were in the soil.*'

We are dealing here with a complete theory for the physical structure of plants, a theory that, as we have seen, Bonnet and Buffon had already outlined, and which would no doubt have long-since taken root in science if Gaudiehaud and Albert Dupetit-Thouars had not imagined that each shoot, as an independent plant would have to have roots that accumulated one by one and were the true cause of the increasing size of the plants. Hugo Mohl, Hétet, and Trécul did not take the trouble to demonstrate with proper rigor that these so-called roots did not exist, and those who take only a superficial view of the matter have been led to believe that these eminent observers have overturned the theory of plants adopted by Bonnet, Buffon, and Goethe when, in fact, they only destroyed a controversial interpretation.

---

[2] Linné, *Aménités académiques*, vol. VI, p. 324.

[3] Goethe, *Essai sur la metamorphose des plantes,* propositions 87–90, 1790.

Goethe's idea that leaves and parts of the flower can be thought of as simple modifications of a single organ and his view of the plant as the result of the complex union of an indefinite number of simpler entities comes straight from a more durable idea, namely that of contriving an ideal plant from which all those that now exist could reasonably be derived. He wrote to Herder in Naples, 'I am confident that I am on the point of finally penetrating the mystery of the origins and organization of plants... The primitive plant is one of the most singular things on Earth, and nature itself shows that what I say is true. With my model as a key, one will be able to reveal an infinite variety of new plants that, if they do not already exist, could exist, and, far from being the reflection of an artistic and poetic imagination, will have a real and even necessary existence. *It will be possible to apply this creative law to all kinds of living things.*'

Goethe evidently envisaged for plants something analogous to what Geoffroy Saint-Hilaire called the uniform anatomical plan of animals. He even extended his idea to animals and his earliest zoological work shows that before concerning himself with botany he sought to find in animal life the same unity that he later saw in plants. It was in this way that, after 1786, he went on to discover in man the two intermaxillary bones that, in all mammals, carry the upper incisors and that had been thought to be a distinctive attribute of humans and apes. Like Geoffroy Saint-Hilaire, it is from studies of fetuses and abnormalities [monstres] that Goethe succeeded in establishing the existence of the two bones that in man are usually welded at an early stage with the two halves of the upper jaw between which they are included and produce, when they remain separated, the deformity known as harelip.[4]

In 1790, the same year when he published his essay on the metamorphosis of plants, Goethe was walking through a Jewish cemetery in Venice and accidentally kicked a sheep's skull breaking it into pieces. These few pieces gave him the idea that the cranium is formed from a certain number of vertebrae that have been modified in their form and proportions. This idea, which Frank and Oken arrived at independently but with completely different conclusions, introduced into comparative anatomy the important idea that the same organ, in repeating and modifying itself, could form parts of an organism with a very different appearance. After much argument, it is now considered pointless to try to determine how many vertebrae make up the cranium, but at least it is no longer questioned that the cranium is only a modified part of the spinal column in which the vertebrae have been enlarged, transformed, and partly welded to make a protective case for the brain.[5]

The discovery of the intermaxillary bone and the recognition that the cranium is formed from vertebrae are elements of a greater, incomparable work resulting from a brilliant line of research that Goethe started in 1795. He set out to construct an ideal plant from which all others could be derived by simple modifications

---

[4] On the existence of an intermaxillary bone in the upper jaws of man and animals *Acta naturæ curiosorum*, vol. XV, 1786.

[5] This idea is no longer generally accepted. (Trans. note)

of certain parts, and he decided to study the skeleton in order 'to establish an anatomical type that could serve as a model for future descriptions of the bones of other sorts of animals. This model should take into account as many physiological functions as possible. No single animal could be taken as the standard type, for it would not reflect a true image of the whole. Man, whose organization is so perfect, could not, by reason of this same perfection, be the basis of comparisons with the lower animals. On the contrary, it was necessary to take a different approach: observations tell us which parts are common to all animals and which differ from one to the other. The appropriate assemblage is then used to deduce a hypothetical form that could serve as a model for various types of natural creatures.'

The same year, Goethe and Geoffroy Saint-Hilaire conceived, each in his own way, the idea of a uniform anatomical plan for the entire animal kingdom. But Geoffroy Saint-Hilaire was able to draw on his extensive research to furnish examples for his idea, while Goethe, after starting to construct his plan from osteological observations, stopped mid-way without drawing significant conclusions from his observations. Like Geoffroy, he chose to determine the nature of organs from their relative positions, but, unfortunately, he attached more importance to their functions. Also like Geoffroy, he explained the reduction of size of certain parts of the body as the result of an excessive development of other parts, but the two naturalists arrived at these ideas by completely independent means.

Goethe added to Geoffroy Saint-Hilaire's hypothesis the idea that an organ from a given animal can undergo various kinds of metamorphoses and attain its definitive form only after passing through a number of successive transformations. In doing this, Goethe made a distinction between plants and animals. The parts that are metamorphosed in plants remain united to them; it is only the last of these parts that are born from one another that takes on a new appearance, but they co-exist with those that have not been metamorphosed. When an insect, for example, metamorphoses itself, it preserves no link with the form it has just abandoned. It is the totality of its being that takes on a new aspect. We shall soon see that this difference is only apparent and that there are animals among which the transformations that Goethe depicted so well for plants are found with all their earlier characteristics preserved in the new form.

Naturally, these metamorphoses caused Goethe to realize that the forms of living creatures are not immutable and that their characteristics could be modified over the course of time. Thus, Goethe, like Lamarck and Geoffroy Saint-Hilaire, became a *transformist*, and he contributed a better appreciation of the influence of ecological factors in determining how organisms are modified.

Although he left little written work, Kielmeyer shared these ideas, and through his teaching, he exercised a strong influence on the minds of German naturalists. We know next to nothing about the man except what we find in a lecture he delivered in 1796 as part of a course he gave at the university of Tübingen. Like Goethe, Kielmeyer met Geoffroy on several occasions, but this does not mean that their ideas were not their own. Kielmeyer, in particular, thought that the lower animals represent the transitional stages that higher animals passed through before reaching their definitive forms. Each lower form can thus be considered the arrested development of a higher one, and in the course of developing, each higher form has passed through earlier ones that are analogous to those preserved in the less developed members of the group to which they all belong. Just as frogs start as true fish, mammals went through a stage when they were reptiles. In 1806 Geoffroy Saint-Hilaire drew attention to the importance of a remark made by Autenrieth in 1800 when he pointed out that mammals have the same number of bones in the skull as fish and other progenitors.

On several occasions, we have encountered an idea that was developed by Serres but would not take on its full philosophical importance until transformism became a recognized scientific concept and was translated into the fundamental proposition that the embryology of an animal is only an abbreviated repetition of the phases that a species has gone through to arrive at its present form.

Correlations of this kind between the lower and higher members of the animal kingdom obviously assume that all the forms are products of a more or less advanced stage in the development of a single plan. The uniform plan of composition had many resolute supporters in Germany as well as France, and the dates of the earliest publications dealing with the idea show that it developed in the two countries simultaneously.

A similar accord between scientists and theorists shows that they shared an idea that, from the moment it was conceived, was in harmony with most of the emerging evidence that attracted their attention. But, as Cuvier maintained, the factual evidence was incomplete. Geoffroy Saint-Hilaire could be accused - and possibly Goethe and Kielmeyer as well - of having gone too far in drawing sweeping conclusions from the valid ideas that they had introduced. But is it really wrong to do that? What one calls – a bit disdainfully – an idea in the natural sciences would be called a rule in the other sciences. The essence of a rule is to coordinate the greatest possible number of known phenomena, but there is almost always a temptation to make it too general. It is the new work that it stimulates that determines the extent of its application. And yet the rule, even when constrained, has a value of its own. It finds its natural place among the consequences of other more general rules and becomes a corollary of a more general truth that will later evolve from it. Thus, by a happy combination of factual observations and rules derived from them, the human mind progresses confidently toward the conquest of higher and higher orders of knowledge, always aspiring to the final truths that could explain the origins and future of mankind.

The impassioned struggles to which the uniform plan of composition led should have had the effect of encouraging independent thinkers to search for more general formulas that could reconcile the two opposed doctrines. Two men tried to achieve this by borrowing some of Goethe's ideas: Richard Owen in England and Dugès in France. The former brought to his studies the same precision that characterized the work of Cuvier, and immediately attracted numerous proselytes. The second, an enthusiastic and perseverant scholar like Geoffroy, died without seeing his work gain the appreciation it deserved, even in his own country.

Chapter XIII

Dugès

Attempts to reconcile the ideas of Cuvier and Geoffroy Saint-Hilaire – Organic conformity in the scale of animals life – Moquin-Tandon and the zoonite theory – Dugès' generalizations of this theory – Theory of the constitution of organisms: law of modification and complication, law of coalescence – Dugès' ideas about the types of organisms.

At the same time that the great academic debate over the uniform anatomical plan of animals was about to be closed by the death of Cuvier, a young professor on the Faculty of Sciences at Montpelier, Antoine Dugès, tried to find a middle ground where he hoped the two camps could come together. Although he was strongly influenced by the ideas of Geoffroy Saint-Hilaire, Dugès was also impressed by the objections that Cuvier had raised. He wondered whether, by slightly modifying the formulation of Geoffroy's basic idea, it might not be possible to save it from the curses that the so-called school of facts was trying to place on it. He was convinced that Cuvier's death had not brought an end to his way of thinking. In the Preface of his *Mémoire sur la conformité organique dans le règne animal*, he said: 'I have decided to publish this memoir in the hope of ending the controversy that began with the nomination of a commission of inquiry by the Academy of Sciences and was suspended only when, for fear of offending him, the Academy charged M. Cuvier with rendering its report. I feared that, under such circumstances he would find it difficult to be impartial and would inhibit the lively discussions in which he expressed his opposition to principles that are very similar to my own. I have tried to avoid any appearance of taking sides in this great quarrel and have only stated personal opinions that I have arrived at independently. I have tried to be fair in citing other views, but I have been unable to moderate the strong feelings that were evident in Cuvier's studies and could not temper the disgust he expressed when referring to any generalizations that he considered too sweeping and hasty. He displayed his strong feelings to me personally, and I doubt whether I was able to soften the harsh attitude that became so apparent during our long conversation.' Dugès gave up trying to support the principle of a uniform plan of composition of the animal kingdom. He assumed only that the different types of animals are linked to one another by systematic modifications. The conformity of a particular pair of animals to such a relationship can be recognized sooner or later, regardless of the class to which they may belong.

A. McBirney, S. Cook and G. Retallack, *The Philosophy of Zoology Before Darwin: A translated and annotated version of the French text by Edmond Perrier*,

What is the nature of this conformity that Dugès substitutes for the uniform plan of animal composition? It would be helpful if Dugès had been more specific. Making allowances for the obscurity and conceptual errors that scientific terminology imposed on him, one can see a promising general idea that, had it been more fully developed, might have had consequences even down to the present day.

Science was just beginning to appreciate the beauty of Goethe's way of viewing the composite nature of plants and the metamorphosis of their organs. Dunal had speculated that there might be something analogous in the animal kingdom and that invertebrates might be assemblages of simpler animals that are grouped together in diverse ways. In 1827, Moquin-Tandon, in his *Monographie des hirudinées*, had brought more precision to this idea by showing that each segment of the body of a leech is identical to those that preceded and follow it and that each of these segments contains all that is needed to live independently as a distinct organism. He considered this unit to be a miniature animal, or *zoonite*. All of Cuvier's articulated animals could be broken down in a similar way and are nothing more than assemblages of simpler animals, zoonites, disposed in linear series. Generalizing on this idea, Dugès tried to show that it applies not only to articulates but to all invertebrates and vertebrates. The polyps of a colony of coral or bryozoa, are zoonites with the same relations as the segments of an insect; the only difference is that they are disposed in different ways. Zoonites can group themselves in linear series or they may be arranged radially around a central point. They can also form branches like the limbs of a plant. Various transitional forms of these different structural arrangements might form links between animals that at first appear to be completely unrelated. Individual zoonites always have the same structure and composition, and the animals they form differ only in the number and arrangement of their constituent parts.

This relationship allowed for an infinite number of intermediate forms, and there could be no sharp demarcation between the different members of the animal kingdom. Dugès hoped that he had discovered the laws governing the uniform plan of composition that Geoffroy had been searching for and that he was making due allowance for the objections raised by Cuvier.

These laws were four in number:

1. *Law of multiplicity of organisms;*
2. *Law of disposition;*
3. *Law of modification and complication;*
4. *Law of coalescence.*

One can state these as follows:

1. All higher animals are composed of a certain number or simpler organisms, known as zoonites.
2. The zoonites making up an animal can group themselves in a number of ways: as a single linear series consisting of two alternating or symmetrical chains, as a layer around a central origin, or as a completely irregular arrangement.

Within a single animal, two or more of these types of groupings may be combined with one another.

3. The zoonites of an individual animal can take on various forms to share or distribute the necessary work and maintain themselves collectively.
4. The zoonites or the organs that they make up can become partially or totally fused together to the extent that it becomes impossible to discriminate one from another.

All these propositions are very precise. Again, Dugès was clearly expressing the physiologists' perception of the role of the various parts making up an organism. After describing the diverse modifications of the parts of a number of insects, he reached the following conclusion:[1]

'The relationships of sensibility and mobility cause the segments to arrange themselves and distribute their essential functions in a way that most easily achieves their common purpose. This rearrangement enables each part to make a more effective contribution, especially in the case of internal functions. We see that certain segments may centralize and perfect organs that are not present in the other segments. It does this either by abandoning them through a partial coalescence that draws all elements of a similar nature toward a common center or by atrophying and causing an organ that has been rendered useless to disappear from most of the segments while developing more fully in a single segment where its function in the total organism is served more effectively. This reorganization and reciprocal interaction of the individual components is a form of competition aimed at perfecting the general vitality of the total organism, and the association of the organisms is in some ways like that of human society. Civilization brings together many individuals who, through their different contributions, augment the resources and benefits shared by the community as a whole. This is in contrast to a tribe of savages that is reduced to the simplest and crudest way of life. A civilized society affords an image of the *economy* of the highest levels of animal life, such as mammals, while the tape worm is an example of the more primitive way of life. A simple community of the latter kind is partitioned into a limited number of functions just as the animal itself is organized in segments that differ little from one to another.'

Today, physiologists still limit these comparisons, in so far as they apply to the vertebrates, to separate parts of the anatomy. It took astonishing boldness for Dugès to propose an idea that has only recently found strong support. He considered the vertebrates to be segmented animals, formed from zoonites in the same manner as insects, but in which a variety of zoonites are mixed in complex ways, as they are in the spider. The division of the spinal column into identical vertebrae is the most obvious sign of this segmentation of the vertebrates, but there are others.

The spinal column of vertebrates contains the same number of nerve pairs as there are vertebrae. Dugès cited the experiments of Chirac and Legallois who showed that the portion of tissue corresponding to each of the nerve pairs possesses a true

---

[1] *Mémoire sur la conformité organique*, p. 31.

autonomy of its own. This led him to compare the tissue of vertebrates to the chain of the ganglia in segmented animals. It proved that, when one compares different animals or even when one considers animals of the same kind, the individual ganglia making up this chain may merge and become welded to one another or, if they were joined in an earlier, more primitive state, they may become separated. Blanchard's research has shown that this first case is more common among insects, but Swammerdam had already shown that the closely associated and almost welded ganglions of the larva of the rhinoceros beetle and cameleon flea separate when the insect arrives at the adult stage. These results have been greatly extended by the work of Künckel d'Herculai and Brandt.

Each vertebra in the dorsal region carries a pair of lateral appendices. These are the seven vertebrae of the cervical region. In the case of mammals they are not present in the five vertebrae of the lumbar region. Dugès pointed out that the five pairs of lumbar nerves and the five cervical pairs join respectively in a plexus and then penetrate into the arms and legs, where they are almost the only nerve channels. The number of fingers that terminate the limbs of almost all terrestrial vertebrates is precisely the same number: five. It is therefore legitimate to consider each of our members as the result of the welding of five appendices corresponding respectively to one of the vertebrae segments that furnish the nerves of those members. These appendices are welded together from the center outward and are completely merged only in the first segment of the members. The second includes two bones, the third three, the fourth four, and each of the four others have five. The hyoid bone and the lower jaw are other appendices of the vertebrates that have kept a form close to that of the sides. Finally, Goethe, Oken, and Geoffroy Saint-Hilaire thought that the skull must be formed from a certain number of vertebrae, welded together as fully as are the segments that constitute the heads of insects, and remain distinguishable only by their appendices.

We have here a series of ingeniously developed ideas that have recently been taken up and extended in an interesting article by Dr. Durand de Gros.[2] Geoffroy Saint-Hilaire's work is unquestionably a positive contribution to the doctrine. Unlike his illustrious predecessor, Dugès no longer tried to explain insects by reference to the vertebrates, nor did he try to find the equivalent of the mammal vertebrae in the segments of the bodies of arthropods. The vertebrae and the spinal column were no longer essential parts that had to be found at any cost. Taking up Geoffroy's proposition, Dugès studied the zoonites in the segmented animals where their nature is most clearly displayed. He determined the mode of association of the zoonites and their various parts, and he attempted to find in the vertebrates traces of a fundamental constitution identical to that of the arthropods. The vertebrae and their appendages are the most precise indications of this basic plan. This time, the comparison was placed on a much more practical basis. Unfortunately

[2] *Les origins animals de l'homme*, v. 1 in-8, Germer Bailliere, 1871.

the terms of comparison that were chosen could yield only illusory results. Moreover, one of the propositions on which Dugès based organic conformity is basically incorrect, and the validity of his theory is seriously compromised.

If the arthropods and vertebrates are in fact formed from zoonites, the existence of which the recent discoveries of Semper and Balfour show is no longer in doubt, then this is the limit of their similarity. Dugès, however, set out on the wrong path when he tried to go beyond a comparison of the immediate consequences of this common mode of construction. Though he may not have realized it, his thinking was dominated by the idea of a unified plan of composition. He skillfully modified this idea in order to make it applicable to the higher animals and zoonites. In his view, all the zoonites are identical, and that accounts for the conformity one sees among the animals: 'Although there is no uniform plan on the scale of animals, there is conformity in the way the component elements are always essentially the same, and their disposition does not vary enough to make a clear distinction between the animals that they constitute.'[3]

To find the analogous elements of which Dugès spoke, one must go back to the constituent elements of the tissues and what we call today the *cells* or the *plastids*; Dugès did not look beyond the zoonites. The zoonites of a vertebrate are not at all comparable to those of an arthropod any more than the zoonites or rays of a starfish are comparable to those of a jellyfish. Dugès' preconceived ideas led him to some obviously contrived conclusions. He equated, for example, the mandibles of insects to the upper jaws of vertebrates, and their jaws to the mandibles of the latter. He strayed even farther from reality when he found an argument in favor of his hypothesis for the multiplicity of bones that form the lower jaw of fish. Nevertheless, he showed remarkably good judgment in resisting the enticement presented by the false concept of similar zoonites, and he preserved all the advantages of his method of comparing vertebrates to segmented animals. At the end of his masterful memoir, when he sought to establish a transition from invertebrates to vertebrates, Dugès looked for intermediate types, not between the arthropods and vertebrates but between the vertebrates and worms, which is precisely where today's zoologists find them. He was convinced that he saw affinities between leeches and lampreys that are really not as closely related as he wanted to believe: the sucking mouth of the leech cannot be realistically compared to that of the lampreys. The respiratory pouches of these fish are in no way analogous to the lateral pouches of the worm, which are nothing more than kidneys, but Dugès could choose this means of comparison because knowledge of the types being compared at that time was imperfect, and he remained impressed by their general resemblances.

Dugés disregarded the complications introduced by the attempts of Geoffroy Saint-Hilaire and Ampère to compare the internal skeletons of second-order vertebrates to the external skeleton of arthropods and went back to the idea that the vertebrates and the arthropods have opposite arrangements in relation to the ground.

[3] *Mémoire sur la conformité organique*, p. 19.

He insisted that the disposition of organs in a radially symmetrical animal is identical to that of a vertebrate lying on its back, and in this way he arrived at a much more legitimate relationship. He recalled that this reversal is manifested even in the embryo, as was shown by Hérold and Rathke, and is a considerable addition to Geoffroy's list of the animals that have abandoned the normal attitude of their progenitors to adopt a somewhat different one. The sloth, for example, almost always hangs from the branches of a tree with its back toward the ground. Bats and various mite parasites walk on their backs, and in the same way a few insects, like the water beetle, swim on their backs. Among the crustaceans there are the notostracan [*apus*, now trops] and anostracans (*Branchippus*), and, among the molluscs, all the heteropods. Examples among the fish include the mustache catfish (*Pimelodus membranaceus*) and some of the suckerfish (remora). In the case of the latter, the dorsal face normally remains applied against a foreign body and is just like the ventral face of the other fish. But there are other equally remarkable changes of attitude among members of the animal kingdom. Man, among the mammals and the penguins among the birds walk upright in a position exactly perpendicular to that of other vertebrates of their class. The flounders (*pleuronectes*) and the 'amphioxus' (*Branchiostoma*), among the fish, and the pen shell (*Pinna*), oysters, single shells (*Anomia*), and giant clams (*Tridacna*), among the mollusks, rest on their side, while *gammarus* (fresh water shrimp), which are crustaceans, walk on their side and swim equally well on either their back or stomach. Many annelids and certain myriapods can walk without difficulty on either their back or stomach, and there are some that move only backwards. Dugès could also have added that the barnacles and sea squirts spend the greater part of their lives with their head down, which is the normal attitude of all the lamellibranch mollusks, as well as the flying lemur (Galeopithews) and bats that sleep and rest in this position. One must conclude from all this that Geoffroy was correct when he said that parts of the bodies of certain animals may be anatomically normal but occupy varied positions with respect to the ground. When comparing different animals, anatomists should not attach too much importance to the attitude of their bodies.

Dugès often took pleasure in trying to make comparisons between the regions of the bodies of different types of animals by means of criteria that are more rigorous than those normally used in science. In so doing, he gave us the only physiological and morphological definition of the head that can be accepted today: 'It is the part of the body that guides all others, and it has modified parts in the sensory organs and the locomotive appendices used for handling and breaking up their food. This region is composed of several segments, but they have often coalesced to the point that even the closest examination cannot distinguish them individually without resorting to conjectures that always leave some doubt as to the true number of segments.' The comparisons that Dugès attempted to make between the articulates and vertebrates were too detailed and led him down a path that eventually proved to be a dead end.

He was able, however, to illustrate the importance of the ideas he presented in his *Mémoire sur la conformité organique,* and his work proved to be a positive contribution to zoology. Although a great admirer of Lamarck and Geoffroy Saint-Hilaire, Dugès was captivated by Cuvier's seductive ideas and failed to foresee the future importance that transformism would have. He seems never to have wondered about the origins of the animals he studied; it appears that he believed they have always been and would always remain what we see today. He noted that some are reduced to a single zoonite and, in the case of the myriapods, that the zoonites are arranged serially. It did not enter his mind, however, that the simple animals, when reduced to a single zoonite, could be the surviving progenitors of animals formed from multiple zoonites, as Lamarck or any other transformist would surely have noted. He did not consider the causes of the diversity of the zoonites he was trying to assign to groups and does not seem to have asked himself whether they developed in parallel or serially or how one could account for the development of what Cuvier called organic types. Quite the contrary, he saw them as primitive forms. He believed that, from the beginning, each animal carried the mark of the type to which it belonged: 'Even in its most primitive form, each species has its own particular form and style, both external and internal, but we cannot define the nature of the inherent design that gives the animal its distinctive mark. All one can say is that it constrains the ways in which a species can propagate and prevents it from losing its distinctive attributes. There must be some sort of power governing the creature's form. Its effects are there for us to see, and it should be possible to study them. The embryo passes through a series of transformations that are analogous to steps in the scale of animals, and it does so without losing its distinctive characteristics.' One can recognize the influence of Von Baer's thinking in these words, but this was in 1831, at a time when the fundamentals of embryology had scarcely begun to emerge. Next to nothing was known about the evolution of the lower animals, and even less was known about the development of the higher ones. Dugès was already in advance of his time when he described the reproduction by transverse division of the *Catenula lemnæ,* a species of planaria (flat worms).

The law of organic conformity resembles that of a uniform plan of composition in that it does not pretend to explain the linkage of animals: it simply defines their common structural elements without attempting to establish a purely theoretical relationship. I am tempted to say purely 'theological.' One feels that his interpretations rested on his fundamental faith and on ideas that were even more metaphysical. At times he reveals a kind of Pythagorean compulsion to show that animals that are otherwise very different always have the same parts in the same number without offering any reason why the preferred number should be constant. For example, Dugès attempted to show that the necks of vertebrates are formed from three vertebrae, just as the thorax of insects has three parts, and he thinks he sees a correspondence between the five pairs of feet of the crustacean decapods and the five primitive appendices which, when welded together, constitute what he considers to be the limbs of the higher vertebrates.

In short, he imagined that the same parts must always be found in the same number and that they can be designated by the same names in both the vertebrates and articulates. He set up a table comparing the parts of the bodies of these animals and arrived at what appeared to be a clear demonstration of their structural identity. Dugès seemed to be convinced that these numerical laws govern the entire animal kingdom. His knowledge was too extensive for him not to have seen the possibility of finding in an insect like the siphonophore all the zoonites of a shrimp or cat, but when one is searching for resemblances that can be explained only by appealing to a superior will, there is no limit to what one can postulate. The numbers have some sort of fundamental significance that certain minds have always found intriguing. Is it not true that MacLeay, a distinguished naturalist, built an entire system of zoological divisions on the importance of the number five which he thought had governed all organic evolution?

This is the same metaphysical tendency that led Dugès to think that the divisions of the animal kingdom can be illustrated by two intersecting circles, one including the invertebrates and the other vertebrates. As can be seen from the illustration shown on the next page, these circles were ingeniously constructed, but they correspond to nothing in nature. Such attempts simply illustrate the author's deep conviction that the continuity of the universe can be expressed by a simple geometrical figure; when Bonnet's straight line failed, Dugès adopted the circle.

Despite the inherent defects his words had at the time they were written, one can recognize the value of the morphological ideas Dugès developed in the *Mémoire sur la conformité organique*. Published just when the struggle between Cuvier and Geoffroy Saint-Hilaire was coming to an end, his memoir was not fully appreciated. Few scientists were prepared to appreciate its full impact, and Dugès himself had only dimly perceived its importance. After his death, parts of his memoir were often borrowed by other naturalists, but the work was seldom cited as anything but a scientific curiosity. One must recognize, however, that it had the same importance for animal morphology as Goethe's essay on morphology and metamorphosis had for plants. Some of the discoveries that soon began to emerge provided a striking confirmation of Dugès' views, while others expanded the horizons he had foreseen earlier. But the common thread was lost almost as soon as it was perceived; Dugès' name is seldom heard today even though he could justly be placed beside Cuvier and Geoffroy Saint-Hilaire. I hope that these few lines may serve to repair the thoughtless injustice of zoologists with regard to one of the most eminent naturalists of this century. This injustice was partly a consequence of the difficult situation that prevailed in France as a result of the struggle between Geoffroy Saint-Hilaire and Cuvier and the discredit that was cast on all the attempts at philosophical syntheses by the excesses of the school of German naturalists that we consider in the next chapter.

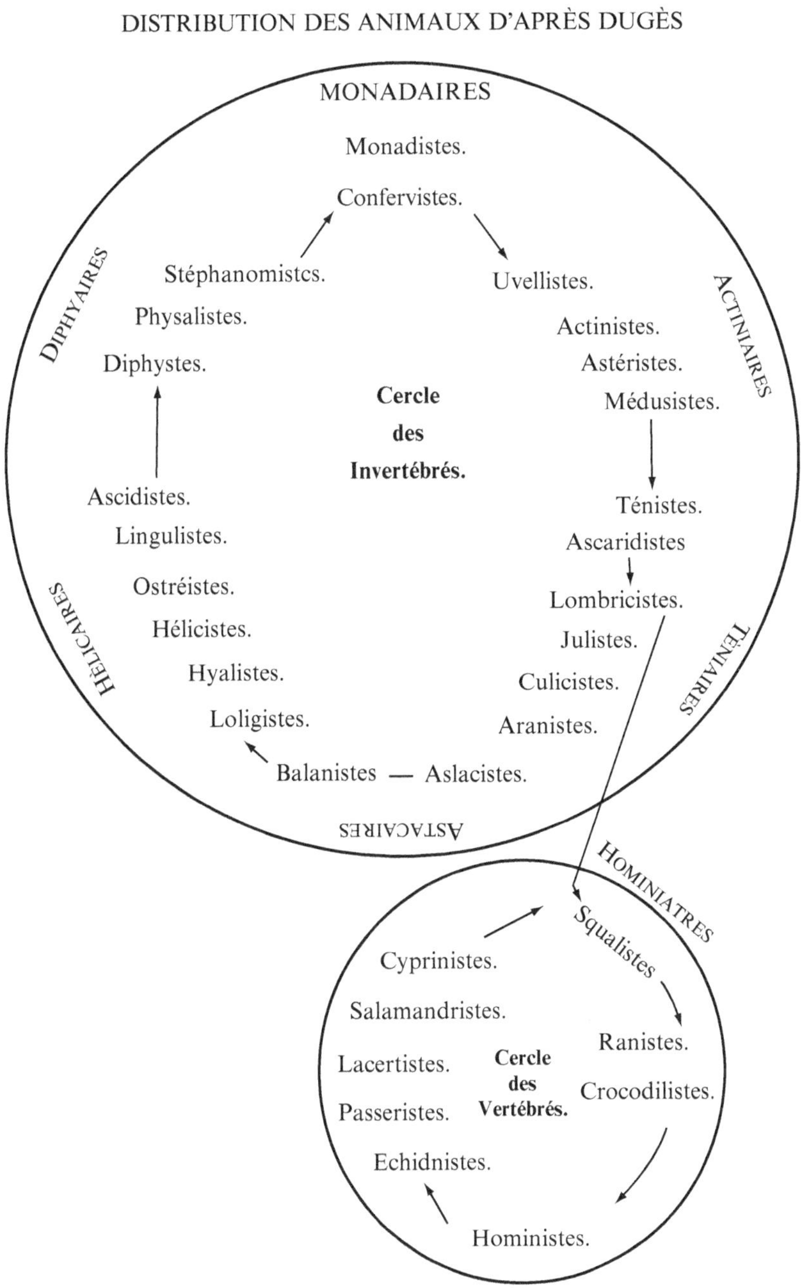

Fig. 13.1 Distribution of animal life according to Dugès

Translator's explanation of Dugès' "Conspectus of the natural relations among subkingdoms of animals"

Dugès explains his Plate 1 on page 111 of his "Memoire sur la Conformité organique dans l'échelle animale": (page 123 of the present volume):

> "The Synoptic Table of the sub-kingdoms and classes of animals is arranged according to their natural linkages and formed into two circles – one for invertebrates and the other for vertebrates. In both, the series descend toward the right and ascend on the left, and one moves in this way, first from the simple to the complex, then from the complex to the simple. I have tried to indicate the degree of advancement ["élévation"] of each family proportionately to that which it occupies in the scale of animal beings when they are considered relative to their gradual perfection and assumed descent ["en la supposant descendante"]. Thus, the monads occupy the first stage, humans the last; molluscs or hélicaires are at almost the same height as the arthropods or astacaires. The salamanders and frogs, batraciens in usual nomenclature, are also at the same level. It is the same with the crocodiles and lizards - the saurians of those authors who would have us separate them out of the lineage to which nature seems to lead our steps, regardless of their relations or their very real resemblances, but nevertheless quite remote from a common ["immédiate"] parent."

Several points may be made about the diagram. The groupings are arranged according to increasing and then decreasing complexity within the circles. As he said, "In departing from the simplest, the Monadaires, one might as readily trace to the right or left, to the Actiniaires or the Diphyaires....[p 77]." On the contrary, the transition from invertebrates to vertebrates is uni-directional, all the latter being more complex than the former.

The second point is that there is no axis of time; this is no dendrogram or family tree.

However (third), it seems clear enough that he seeks natural relations among the classes as he looks for transitional groups: "at the point of union between monadiaires and actiniaires (radiaires) there is a little class of beings, which I call uvellistes. These contain animals that are composed of a group of loosely attached monads [cells] ...." He exemplifies the Uvellistes with two species of *Volvox* and *Gonium pectorale* - genera in the Volvocales. In similar fashion, he discusses transitions between other groups within and between the invertebrate and vertebrate circles. The subjects of these discussions are indicated by the arrows of his diagram.

Dugès named the classes in his taxonomic hierarchy in a systematic way by choosing nomenclatural type genera in the Linnean fashion and by giving a unique suffix to each rank above the genus (p 102 of "Memoire..."). Thus, he designates a family by "é," tribe by "ique," order by "ien," class by "iste" and subkingdom by "aire."

Unfortunately, in Perrier's rendition of Dugès' original diagram, the ranking of classes under subkingdoms is misleading: the word "Téniaires" should be placed along side of only "Ténistes" and "Ascaridistes." And Astacaires includes all classes from Lombricistes to Balanistes. Indeed, the clearest distribution of the classes into subkingdoms is indicated by the arrows of the diagram.

Dugès, in presenting his classes and subkingdoms on pp 104–110 of the "Memoire.,.," gives us a convenient framework for translating the epithets of the diagram of circles of affinities:

The first subkingdom consists of the "Monadaires," "Animals of a single cell and of very simple organization" (Dugès, p 104). The class, Monadistes, is based on the genus Monas. A monad is a single-celled flagellate protozoan. "In early commentaries, such organisms are often regarded as the lowest members of the hierarchy of nature or as fundamental units of the animal body" (Oxford English Dictionary). Conferva, the type genus of the class Confervistes, was "A genus of plants originally constituted by Dillenius [yr 1742], and then made to contain many heterogeneous species of filamentous cryptogams" (Oxford English Dictionary). The word goes back to Pliny.

The second subkingdom is made up of "Actiniaires" or "Radiaires." "These are composite [multicellular] animals, aggregated or bound together in a circular symmetry." They are actinomorphous or radial in symmetry. The class Uvellistes includes two families, one of organisms formed by undifferentiated cells aggregated into spheres (cf. *Volvox*) and the other aggregated into sheets (cf. *Gonium*). Dugès seems to have coined the name of this class (p 78) without basing it on a genus but, rather, with reference to the Latin word for grape, "uva." Actinistes (or polyps) include the family of hydras and five other families. The class Asteristes (echinoderms) has four families. The class Medusiste includes jellyfish (Coelenterata) and comb jellies (Ctenophora, vis. Beroe).

The third subkingdom is that of the "Téniaires." The class, Téniste includes (among three families) the tapeworms of the platyhelminthes. The class, Ascaridistes includes in its first order, roundworms, annelid worms, and proboscis worms. Its second order includes flatworms and flukes.

The fourth subkingdom consists of the "Astacaires" or articulates, animals with segmented bodies. This large group is composed of six classes: Lombricistes, annelid worms; Julistes, the myriapods - centipedes and millipedes; Culicistes, insects; Aranistes, arachnids; Astacistes, crustacea – crayfish; Balanistes, cirripedia or barnacles.

The fifth subkingdom comprises the "Hélicaires" or molluscs. The name is from the type genus *Helix*, the European garden snail. There are six classes: Loligistes, cephalopods, squids and octopuses; Hyalistes, pteropod molluscs or sea butterflies; Helicistes, gastropod molluscs, snails, slugs; Ostréistes, bivalve molluscs; Lingulistes, the modern phylum, brachiopoda, lampshells; Ascidistes, the modern subphylum Urochordata, tunicates and salps.

The sixth subkingdom is composed of "Diphyaires." The three classes of colonial jellyfishes are all members of the hydrozoa of the order Siphonophorae. They take their names from the following genera: *Diphyes* of Chamisso, 1821; *Physalia* (Portuguese man of war) of Linnaeus, 1758; and *Stephanomia* of Peron, 1807.

The seventh subkingdom is that of "Hominiaires." It includes nine classes of vertebrates: Squalistes, cartilaginous fish; Ranistes, anuran amphibians, frogs and toads; Crocodilistes, testudines (turtles and tortoises) and crocodiles; Hoministes, mammals, with nine orders; Échidnistes, monotremes (cf. *Platypus*); Passéristes, birds; Lacertistes, lizards and snakes; Salamandristes, urodele amphibians, newts and salamanders; Cyprinistes, bony fishes.

Chapter XIV

The German Natural Philosophers

Schelling's ideas. – Oken: polarities and the origin of the universe. – The primitive mucus. – Equivocal generation of the infusoria, anatomical elements. – The law of repetition deduced from natural philosophy. – Man and the microcosm. – Degrees of organization. – Theory of the vertebrates; constitution of the cranium as vertebrae. – Spix: application of the law of repetition to comparative anatomy. – Carus: Extension of the theory of vertebrates.

The great school that began with Buffon and continued with Lamarck, Geoffroy Saint-Hilaire, and Dugès in France and with Goethe and Kielmeyer in Germany, assembled large amounts of factual information and through a series of inductions, tried to use these facts to derive a general concept relating the various forms of life. It was hoped that this would lead to discoveries of new facts and relationships. Their approach was one that is commonly followed in many realms of science, and their methods were unusual only in the great mass of factual information that they used in their attempts to find broad, general relationships. Later philosophers chose a rather different approach: starting with a sweeping, abstract idea, they used a preconceived concept to deduce facts by pure reason. This is what was attempted in Germany at the beginning of this century by what is known as the school of *naturphilosophie*.

It would seem that an approach of this kind would necessarily be sterile, but this is not always the case. No matter how they are expressed, all ideas have some sort of factual basis, and they always contain some element of reality. For one thing, the consequences of a theory cannot be explored without keeping in mind the facts that it is meant to explain. The mind can never be at rest until, in one way or another, it is able to make the facts consistent to some degree with the principal idea, but each time one has recourse to this procedure the basic concept is modified a bit and with time becomes increasingly realistic. What began as an attempt to relate abstract ideas gradually becomes an effort to establish relationships among the facts on which those ideas are based, and from these perceived relationships one gains a better appreciation of their true significance. This was the case with the school of Natural Philosophers, and it explains the enthusiasm that their work aroused and the influence it exercised in Germany for nearly half a century. It also explains how a movement that at first seemed irrational was ultimately successful and made important new discoveries.

A. McBirney, S. Cook and G. Retallack, *The Philosophy of Zoology Before Darwin: A translated and annotated version of the French text by Edmond Perrier*,

The foremost of the natural philosophers, Schelling, had followed the lessons he learned from Kielmeyer and found ways to incorporate all his illustrious master's ideas into a system[1] based on certain forces and beings, often hypothetical, that seem to neutralize each other by their union. For example, negative and positive electrical charges are both active forms of energy, but when combined, they produce a simple electrical current with no visible manifestation. Similarly, the flux between two magnetic poles, the boreal and austral fields, neutralize each other in the same way. The two sexes of animals and plants, when isolated from one another, are incapable of producing anything, but their union yields something tangible that defines their species. Schelling eventually concluded that this opposition, whether real or only apparent, is the ultimate general law by which all things are governed. Of all the many kinds of opposite pairs, the most general are in the self and others, *unity* and *plurality*, the *mental* and *material worlds*. These oppositions, like electrical polarity, are only different manifestations of a universal principle that Schelling called the *absolute*.

When the human physical and mental identities are opposed to one another, they act like the two electrical polarities and ceaselessly tend to unite, but once united they are neutralized and become inert. In their attraction to one another, these two forces must overcome obstacles, just as an electrical current must overcome resistance, and it is these obstacles that constitute all the tangible features of the world and the creatures that inhabit it. By itself, an electrical current has no visible manifestation; it reveals its existence when it encounters resistance and yields its energy in the form of heat, light, or mechanical force. In the same way, the mental and physical elements of a living being are two parts of the whole, the mind that conceives an action and the body that carries it out. One might say that, in a sense, the two parts of the whole, the mind and the body, have created the world and that one need only look at oneself to find all the essential components of this system. It reminds us of the famous aphorism: 'To philosophize on nature is to create nature.'

Living beings are nothing but successive stages of a single, greater activity. As Kielmeyer maintained, the highest orders of life have evolved to their present state by adapting to obstacles that simpler forms were unable to surmount. Their organs must have developed from those of lower forms. This is in accord with the doctrine of epigenesis that was an insurmountable obstacle for Buffon. All things, both organic and inorganic, are nothing more than manifestations of the same general activity in which all living things participate. The entire universe is nothing more than an immense organism with physical and spiritual parts that constitute the absolute being, namely God, which would be nothingness if the world did not exist.

In developing this system Schelling deliberately confined himself to generalities, but Lorenz Oken undertook to delve into the minute details of the phenomena, and in doing so he gave it a broader scope by drawing on mathematics and the physical

---

[1] See: *De l'ame du monde, hypothèse de haute physique pour expliquer l'organisme universel*, 1798, and *Premier plan d'un system de philosophie de la nature*, 1799.

and biological sciences to arrive at more rigorous arguments and comparisons. All his philosophy rests on this identity:

$$+A - A = 0$$

which is a mathematical expression of Schelling's general principle of opposition or polarization. This simple equation contains both the material universe represented by the term +A, and the realms of the mind, –A. The intimate union of these two terms is the divine, the absolute; it is zero, the nothingness from which all has immerged. The material universe, finite space and time, is the passive absolute; the ideal, infinite, and eternal is the active absolute. In opposing the active and passive in this way, the absolute achieves creation. The active absolute that *proposes* and the passive absolute that *responds* to the proposal come together in a *unification* just as the plus and minus are combined in zero. These three forms of the absolute are the three persons of the Trinity, which is God. Oken thought he saw in them the explanation of many other mysteries, but he did not always dwell on these sublime heights. He descended to establish a principle that is rather similar to the mechanical principle of *action* and *reaction*. According to him, all force is double and composed of a negative and positive components. Motion results from this polarization of force, in which the two terms always tend to neutralize themselves without ever completely doing so. The more numerous and different the opposed terms composing a single force, the stronger the motion they generate. But motion is life, and the greater the contrast and diversity of its components the stronger it will be. The strongest form of life is now man: he contains all these diversities, and each of these diversities is a possible form of life, a being. Thus man incorporates in himself the entire world. Every animal is only a reduced form of man, an isolated organ or an assemblage of a certain number of organs that are found in man. This concept is the basis of a whole zoological system that we shall explore shortly.

But how have living creatures been formed? To explain this, we must first examine the various parts that Oken's system has linked together as rigorously as a series of theorems in geometry.

The absolute creates matter by combining two opposites. Matter, being only the passive absolute, is a single entity: the *ether*. The unpolarized absolute corresponds to zero and is represented by a point; the extended point expands by repulsive forces and becomes a sphere. Thus the ether is spherical. It tends to re-enter the absolute and gravitate toward the center; it is heavy and always in motion, but it cannot unite with the absolute; it circles around it. The absolute is the point, the center; the entire sphere turns around its center.[2]

---

[2] Oken's concepts bear an uncanny resemblance to certain elements of modern physics in which the universe is viewed as a system of positive and negative particles or as matter and anti-matter that combine to create the various forms of matter and energy. Note the similarity to black holes and curved space in the paragraphs that follow. (Trans. note)

Like the absolute, the ether has a dual nature. It must become polarized. It can do so only by dividing itself, like the absolute, into rotating spheres, some active, others passive. Polarized in this way, the ether gives rise to the galaxies; the active spheres are the suns, the passive ones are the planets that are pulled toward the suns in order to re-enter their absolute and, as a consequence, are turning around each other. The force that separates the suns from the planets is what we see as light; this force causes the ether to be polarized into suns and planets. It is produced at the expense of the ether, the material of the physicists. Thus there can be no matter without light. Heat is born from the struggle of light with the unpolarized ether; light and heat combine to make fire.

The planets, like suns, are a *trinity*, an absolute of which the active and passive elements, the liquids and solids, are separated by a force [tension] in the intervening gases of the air. Oken used the term *galvanism* to designate the assemblage of these three parts, solid, liquid, and gas. Minerals, which are one of the products of this polarization, owe their solidity to a new bonding force, a kind of *magnetic attraction* of their component parts that manifests itself in their crystalline form. Heat ionizes the crystals. Another force of a more chemical nature comes from the neutralization of the two polarized charges, and this disassociating force tends to change the crystalline solid into a liquid.

Chemical reactions have transformed mineral matter into new states, the highest of which is a form of carbon. As this carbon passed through the successive stages of liquefaction and vaporization or oxidation that constitute galvanism, the solid, liquid, and plastic forms became an amorphous substance, the *primitive gel* or *Urschleim*. The gel and salt were distributed uniformly in the sea and became polarized by sunlight. When the gel in the sea became organized, life began to emerged from it. Life is only a form of galvanism. The primitive gel had to have the three essential powers of solidification, liquification, and oxidation that correspond to the three functions of assimilation, digestion, and respiration. In this way the primitive gel was able to organize itself, just as the primitive ether did. Not being able to form a single sphere that would reconstitute the planet, it divided itself into an infinite number of small spheres. These spheres are the infusoria that were born directly from the gel by *univocal generation.* Plants and animals are nothing but agglomerations of infusoria. In dissociating they resolve themselves into an infinite number of infusoria that appear by *equivocal generation.*

It was the action of light that brought about the transformations of the infusoria into plants and animals. Plants remained attached to the ground because they had not been sufficiently affected by the action of light. They thrust themselves up from the soil in search of light, and when sufficiently exposed they produced blossoms, but they still remained tied to the earth as the earth is to the sun. In the trinity that makes up the living world, they represent the planetary element, while animals, free like the sun, which is bound to nothing, are the solar counterpart. Plants contain components that correspond to three features of the planet, solidity, moisture, and elasticity; animals have, in addition to these three, a fourth that represents solar light. This basic element is already present in the most noble part of the plant, the flower, which has evolved back to the origin of everything, to the *point source* represented by grains of pollen. The animal is a flower without a stem. It begins where the plant ends. At first, it was only a kind of seed animated by light, a 'sensitive uterus.' This is the case of infusoria. All the parts of the plant are represented in the animal but are ennobled by light. The animal itself is a system analogous to the cosmic system. The planets are represented by their bones and the sun by a nervous system formed from grains of pollen but united to form a single unit. The flesh is a medial part that partakes of both the bones and nerves.

Following the same general model and imagining that each stage in the evolution of the world is reached by splitting of an earlier stage into two parts that are held together by their attraction to one another, Oken combined these different terms and came closer and closer to a very detailed representation of all phenomena. Each object and each phenomenon, being drawn from a pre-existing object or phenomenon and able to produce new objects and phenomena by repetition of a single process, it is evident that each of the terms of an evolutionary series is represented in all the others. From this comes the famous aphorism: 'All is in all' of which the *repetition of the parts* in the organism is only one consequence.

As shown elsewhere,[3] this repetitive sequence of parts is a simple consequence of a more general phenomenon, reproduction. The cellular constitution of organisms, epigenesis, the division of the bodies of articulated or striped animals into segments equivalent to one another with fully developed vertebrae as the fundamental basis of the skeleton is the result of observable reproductive processes. A system that is based, as Oken's is, on indefinite repetitions of the same actions must be in accord with all the times that nature has presented real repetitions of this kind. This is precisely the case for plants and animals, as Goethe had correctly concluded from his own observations. He would find himself in accord with nature in all cases in which the opposed influences of two conflicting causes tend to neutralize one another. Thus, observations confirmed certain *à priori* notions of Oken, such as:

'The fixity of species is in large part due to their reproduction by two complementary sexes.

'Plants and animals are composed of elements [cells] that were originally similar to one another and were homologous to those of infusoria.

'All living beings develop by epigenesis.

'Higher organisms result from the union of similar parts that repeat themselves by disposing themselves in diverse ways.

'Many lower organisms can be considered the result of an association of a certain number of organs or parts that are complete only in the higher organisms.'

It is true that some of these truths had already been recognized by others through a quite different approach. Moreover, Oken only explored a small part of the real world that he encountered in the course of a hasty survey. The impact of

---

[3] Perrier 1881 *Colonies animals*, p. 710.

what he found hardly made him pause, and he quickly returned to his unconstrained approach, plunging with renewed ardor into an infinite realm of speculation.

In his studies of animals, he was concerned mainly with identifying the component parts that correspond to those of the group as a whole. The animal, like the cells that make up its body, is only a simple vesicle wrapped in a skin. Initially, the vesicle consisted of nothing more than its skin. The digestive tract of this primitive vesicle is only a part of the skin that has been folded inward and cut off from the effect of light. Under the influence of the atmosphere, the skin develops gills, and the lungs are just gills which have been inverted and placed inside the body. The aorta is a repetition of the windpipe, just as the thoracic duct is. The liver is like a brain to which the intestinal and pulmonary vessels are attached just as the nerves lead to the brain proper. The gall bladder is analogous to the intestines in a system in which the lungs represent the skin; this system having developed in the absence of light, like the fetus which at first is nothing but a liver. The bone system is derived from the liver after that organ was first exposed to light. It shelters the nervous system and serves to sustain the muscular system. The stomach and the back of the animal represent themselves respectively, but the back is exposed to light as the stomach is to the ground, and their orientation is reciprocal. The belly, being only partly exposed to the effects of light, has only a vertebrate column - the sternum. The stomach of the animal represents a part that has remained vegetal. The skeleton also has its animal and vegetal parts: the vertebral disks and ribs that are more animated and joined one to another; a hand results from the merging of five ribs, which are represented by the digits.

The head is the quintessentially animal part of the belly. The trunk, which is already polarized into an animal back and vegetal belly, remains, by nature, more vegetal. It equates to the more elevated art of the plant and represents a sexual animal opposed to the cerebral animal. But the head recapitulates the trunk, for the skull is composed of a vertebral column that can be broken down into vertebrae; the jaws are arms; teeth are digits; the nose is a thorax; the ethmoid bone is a lung; the mouth is a stomach; the soft palate a diaphragm; and the brachial plexuses are legs.[4]

Even more, the skin, intestines, lungs, flesh, nervous system are equally complete beings that represent one another reciprocally. Each is an organism and its complete development culminates in producing one of the organs of the senses, somewhat in the manner of a flower. The flower being equivalent to an animal, each organ of the senses is a parasite in which the entire animal is represented. The most perfect of all is the eye, a true brain that looks outward through the skin.

The sexual animal produces in turn the cerebral animal. This accounts for the resemblance between the anterior and posterior members: the pelvic girdle [bassin] is the thorax of the sexual animal; the ilium its shoulder blade; the ischium its collar bone; the femur its humerus; etc.

---

[4] These analogies are truly the products of Oken's fertile imagination – not Perrier's. He has found in the parts of the head their corresponding parts in the trunk by applying the law of repetition of parts ('la loi de la répetition des parties') and, as Perrier said, Oken 'threw himself into the infinite realm of speculations.' (Trans. note)

It was impossible that in this ardent search for organic repetitions where the most fugitive resemblances serve to justify the strangest correlations, some real similarities of diverse parts of the body could not be fully expressed. Oken joined Vicq-d'Azyr in proposing the homology of the anterior and posterior members, and he followed Goethe in establishing the vertebral constitution of the cranium. He was often able to perceive the essential attributes of an organ; an incisive sentence appears abruptly among his metaphores that signaled an unexpected relationship, and its serious meaning is thereafter inscribed on the mind. Many of these statements and expressions have become part of the current vocabulary of naturalists!

If each part of a man's body is only a miniature repetition of the whole man, the animal kingdom, as we have said, also repeats man as well. Animals are only the organs contained in man, isolated or rearranged in different ways. Thus, animals can be classed according to their degree of complexity, and Oken designated each group by the name of the system that he thought was dominant in it. Here is the table he devised to show these relationships in the animal kingdom:

1st Degree – Intestines, bodies, sense of feeling of animals: Invertebrates
- 1st Cycle. – Animal digestion: radiates
  - Cl. 1. – Stomach animals: infusoria
  - Cl. 2. – Intestine animals: polyps
  - Cl. 3. – Lymphatic animals: jelly fish (medusae)
- 2nd Cycle. – Animal circulation; Molluscs
  - Cl. 4. – Acèphales
  - Cl. 5. – Gastropods
  - Cl. 6. – Cephalopods
- 3rd Cycle. – Animal respiration: Articulates
  - Cl. 7. – Skin animals: worms
  - Cl. 8. – Limb animals: crustaceans
  - Cl. 9. – Nerve animals: insects
- 2nd Degree. – Flesh animals, head animals: vertebrates
- 4th Cycle. – Charnel animals
  - Cl. 10. – Bone animals: fish
  - Cl. 11. – Muscle animals: reptiles
  - Cl. 12. – Nerve animals: birds
- 5th Cycle. – Sensate animals
  - Cl. 13. – Sensate animals: mammals

The same system is naturally followed with rigorous logic in each division. The only place where *à priori* assumptions may play a role is in the naming of the divisions. The delineation of the divisions is in accord with criteria indicated by discoveries that were emerging in the zoological world. Oken simply made these discoveries fit the requirements of his system and did not fail to take account of relevant research. As editor of *Isus*, a journal that is well known for its impartiality, he conscientiously recorded new contributions to all the natural sciences. His studies led him into the fields of osteology and embryogenesis, and through the original ideas he expressed in his teaching and publications, he rapidly acquired an immense influence and provoked a very remarkable scientific movement. He deserves to be placed among those who have rendered a real service to the natural sciences, and even if the basic idea of his system were to collapse, a great many valid ideas, new approaches, and original observations that he made along the way would remain definite additions to the store of human knowledge. His ideas continue to be with us down to the present day. The university of Jena, where he was one of the most eminent professors, remains a leading institution, and the words of his successor, Ernst Hæckel, are often a distant echo of his voice.

Like Oken, Hæckel assigned a dominant role to carbon in the generation of organized bodies. He believed that he had found the primitive gelatinous material, the *Urschleim*, in the famous *Bathybius* which was dredged from the floor of the Atlantic Ocean by the research vessel *Porcupine.* To a large degree the well known and varied theories of the *planula* and the *gastrula* represent rather well the successive phases of development of animals much in the way Oken perceived them. And again like Oken, Hæckel, recognized that certain animals can become arrested in their development while in the state of an isolated organ.[5] Could there be an analogy between this unique process which Oken thought had helped to create the world and the monism that was the basis of Hæckel's philosophy?

It was difficult to exaggerate the importance of Oken's ideas. Contrary to what ordinarily happens, his students made a special effort to restrain the implications of his work and tried to keep them within the bounds of reality, so that one could better judge the true significance of the bizarre fantasies their master had constructed.

Spix (1781–1826) confined himself to saying that nature develops by stages and that each stage is simply a minor improvement of an immediately lower stage. The world began with water, the water became air, and the air became light. One cannot help being a bit confused when eminent men speak of such transformations more than thirty years after the death of Lavoisier when the basis of chemistry had long since been solidly established. This successive development of the parts is manifested more clearly in organic rather than inorganic nature. It culminates in the flowers of plants and with the heads of animals. One might say that the simplest animals (zoophytes and worms) are reduced to an abdomen. In the case of

[5] Clues to understanding this idea of an independent organ may be found in Perrier's earlier description of Oken's belief that animals are no more than imperfect reductions of the perfect animal, the human (Chapter 14): 'Tout animal n'est qu'une réduction de l'homme, un organe isolé, ou un assemblage d'un certain nombre des organs qui se trouvent dans l'homme.' Another expression of this idea is seen in Chapter 15: 'Dans certain animaux' dit en 1827 M. Milne-Edwards, ' le corps présent partout des caractères identiques et ne parait renfermer aucun organe distinct… Les polyps d'eau douce prèsentent une structure de ce genre…' In other words, certain animals are so simple that their cells are alike, i.e. little differentiated – fresh water polyps, or planulas. Hence, they are organs in and of themselves. Perrier may have had these animals and Volvox (the 'Uveliste') in mind when he spoke of the 'organe isolé.' (Trans. note)

fish, the head is just beginning to become distinct; it is better developed in reptiles and birds but has reached its full development only in mammals. The pelvis, the bones of the abdomen, the thorax, and the bones of the chest are only stages in the development of a cephalic skeleton. One finds in the head representatives of all parts of the body, but to identify these representations, Spix, like Geoffroy Saint-Hilaire, Goethe, Autenrieth, and Oken, looked to embryos. He supported his ideas with excellent studies of comparative osteology and embryology, which are basic contributions to science. It is true that this work did not have the same rigorous quality that characterized Geoffroy's approach to problems, but in this case he was dealing with problems that were quite different from those that were of interest to the French scientist. The German natural philosophers were not just comparing one animal to another. As Vicq d'Azyr was the first to show independently of all theory, they compared the different parts within the same animal and tried to find where they share common attributes.

Nevertheless, the studies in Germany and France benefited from a reciprocal influence on one another. For example, in 1824, Geoffroy wanted to determine the vertebral composition of the cranium, and by an ingenious definition of the vertebrae got around most of the difficulties raised by the metaphysical concepts of the German naturalists. Inversely, in 1828, Carus took up Geoffroy's idea that centered the life of articulated animals in their vertebral column. He considered three kinds of vertebrae: primitive vertebrae that form protective walls of the body, secondary vertebrae that protect the nervous system, and a third that separates this system from the rest of the body. Articulated animals possess only the first of these; the vertebrates, in contrast, have three vertebrae combined with one another. For Carus, as with Oken, everything comes down to the vertebrae. The bones of the arms and legs are radial offshoots of the vertebrae. Carus did not confine himself to making anatomical comparisons. He had a whole philosophical system that was only a modification of Oken's. He too attributed all the vital functions to a kind of polarization, and because this polarization is indefinitely repeated he came to the logical conclusion that as the organism develops it simply repeats itself. Thus the rings of the annelids are only repetitions of the first of these, an idea to which Moquib-Tandon was led independently by comparative anatomy. As we have seen, Dugès made brilliant use of this same idea three years later.

If, in considering Carus' comparative anatomy, we avoid the word vertebrae that the disciples of Oken used for all solid parts, and if we also ignore the metaphysical relationships he assumed, there still remain some morphological ideas that have since proved to be quite useful. In particular, one must certainly attach the bones that one finds in the vertebrates to several systems. The most ancient animals possessed a very well developed exoskeleton of which the scales of fish, the bony plates in the hide of crocodiles, and the carapace of the tortoise are some of the diverse modifications. The spinal column developed under the nervous system, and the bones of the limbs make up a completely different system, but, as we see in the tortoise, these two systems can become somewhat mixed. The eminent anatomist, Gegenbauer, recently found that in order to take into account all the peculiarities presented by the diverse forms of the skeleton, he was obliged to call on all the bones of both the external and internal skeleton. Carus explained the existence of these various classes of skeletons by the necessity that the primitive animal, the embryo, has to delimit itself from the external world. A part of the living substance is devoted to the production of this barrier, but at the same time it ceases to live and becomes an inert solid. The animal first covers its external parts by producing a kind of protective shield. Those that remain in this state are 'shelled animals' or 'animal-eggs'. But an animal cannot adjust to the external world without a digestive cavity; it must reconcile itself to this need as well, and this accounts for the other solid parts with which the stomach of so many lower animals are protected. The nervous system of animals that have only these two limits is naturally enclosed in the cavity of the body with the viscera: these are the trunk-animals. But then the nervous system that governs the entire organism also differentiates itself as well. A skeleton is formed around it to provide protection, and in this way the cephalic animals are formed.

Animals with trunks are divided into stomach animals, like the molluscs, and breast animals, like the articulates. Analogous divisions can be found among the vertebrates.

One can see that Carus attached great importance to the nervous system. For him, it was almost an animal within an animal. Oken had no particular ideas about it, and one wonders whether even Cuvier, who had remained in contact with Kielmeyer and his students, had been drawn into the ideas of this school and may have made it play a dominant role in some of his later classifications. However that may be, there is one thing in Carus' work that must be recognized, namely the relationship between the degree of development of the nervous system and that of the skeleton. It is the exceptional development of their nervous system that distinguishes the vertebrates from all other animals, and this in turn has required the development of the spinal cord, which supports it and forms the basis for a vertebral column to which other secondary parts were later added independently.

The anatomical and embryonic research carried out by the German school of natural philosophy and others with similar views inevitably led to a reaction against these exaggerated ideas. The school's influence gradually disappeared, even in Germany. Ehrenberg devoted his entire career to observations of microscopic organisms and said he was glad he had been able to escape the influence of the doctrines that impassioned his compatriots at that time. By his discoveries bearing on the degrees of complexity of animal life, this scientist, who was so well known for his work on infusoria, delivered a terrible blow to the theory of primitive gels (*Urschleim*) and, as a consequence, to the school's entire doctrine. But the factual information and true relationships that these discoveries revealed, together with the new philosophical methods of interpreting the observed phenomena, confirmed the old axiom: "It is through errors that humanity proceeds toward the conquest of truth. Mistakes lead to progress."

This school of natural philosophy had little influence outside of Germany. In France, Cuvier and Geoffroy Saint-Hilaire took a very different approach. Each had his loyal partisans, but there were alliances between the two schools. If the hypothesis of a unified plan of composition, as Geoffroy Saint-Hilaire saw it, fell in the face of facts, the principle of connections[6] has survived, and one can make good use of it in comparisons of the animals that Cuvier assigned to the same major branches on animal life. We tend to forget some of these questions about origins and concentrate all our attention on the determination of the natural relationships of living creatures. We try to deduce all we can from combined views of Cuvier and Geoffroy and utilize their basic scientific contributions.

We now recognize that, in dealing with animals of the same major branches, the physiological organization of the characteristic members commonly used as representatives of the group, are actually rather variable. One tries to determine the limits of variations and construct a common model from which other animals of the same major branch would be only secondary modifications. We would like to discover the philosophical significance of these types, and we have opened a path for naturalists who will soon be wondering about the origin and purpose of these beings that have served as models for so many other animals. It is to this work that we must now turn our attention.

[6] See 'Principle of connections' in the Glossary.

Chapter XV

The Theory of the Organic Types and its Consequences

Richard Owen – the archetypical skeleton. – Analogy, homology, homeotypes. Theory of the vertebrate segment. – The ideal vertebrate; the existence of God. – Owen's views on Transformism. – Savigny: the uniform structure of the mouths of insects. – Audouin; the uniform structure of skeletons of articulated animals. – H. Milne-Edwards and the typical articulated animal; fundamental identity of the zoonites; significance of the regions of the body; law of the division of physiological labor and its importance. – The growth of the body and agamic reproduction of the articulates; identity of these two phenomena; the significance of zoonites. – Parallels between the laws of the constitution of animals and the laws of political economics. – Studies of the lower animals: de Quatrefages, Blanchard, and de Lacaze-Duthiers.

The studies of Geoffroy Saint-Hilaire, the brilliant insights of Goethe, and the speculations of the German school of Natural Philosophy drew renewed attention to the various orders of resemblance of the different vertebrate animals. Thanks to the new ideas about the skeleton that emerged from these studies, osteology soon took on the status of a true science. It seemed that bones – solid and invariable in their appearance and relative positions – were a basic feature to which all the organic systems could be related. Bones have determined the arrangement of organs, and if the vertebrates really shared a definite plan of composition, it was in the skeleton that one should find it best displayed. Goethe had recommended a methodical and relentless pursuit of this study in the hope that it would enable one to identify a general type that could serve as a standard to which all the varied skeletons of the animals could be related. This is the problem that Richard Owen undertook to resolve. He coined the name *archetype* for the primordial skeleton to which he hoped to be able to relate all the others.[1]

---

[1] Richard Owen's first theoretical views on the constitution of the skeleton appeared in 1838 (*Geological Transactions*, p. 518). His *Principles of Comparative Osteology*, (1848) published in French in 1855, as *Principes d'osteologie comparée,* and his *Lectures on physiology and comparative anatomy of the vertebrates* are more complete accounts of his work.

One can do this only by selecting several clearly defined sets for comparisons. In this respect, the comparisons of different forms of vertebrates that Geoffroy Saint-Hilaire had used so successfully naturally come to mind. The most obvious observation that emerges from such comparisons seems to be that most of the vertebrates have organs designed to accomplish the same basic functions. All the species possess appropriate organs for filling the same purposes: eyes, ears, mouth, and a digestive track. They have limbs that are used as legs by vertebrates that walk, as wings by those that fly, and as fins by those that swim. Thus the word analog did not have the same meaning for Owen as it did for Geoffroy Saint-Hilaire who used it for organs occupying an identical position in closely related animals with similar anatomical structures and embryonic origins, even though the organs may have very diverse functions. These organs, which, in any proper anatomical terminology, should bear the same name, were designated by Owen as *homologues*. In order to illustrate the difference between analogous and homologous organs, he cited the small flying dragon, a remarkable reptile that possesses both feet and wings. These wings, which serve to sustain it in the air, have the same function as those of birds and are analogous to them, but they have an anatomical composition that is quite different from that of other appendages and are therefore not homologous. On the contrary, the fore feet of the same dragon have a structural relationship that is clearly similar to that of the wings of birds. Although these organs serve different functions, some for walking and others for flying, they are no less homologous. Like Geoffroy, Owen made the way in which organs are connected to the body his main criterion for designating them as homologous.

These homologous organs are the only ones that should be used for comparisons when one wants to arrive at a determination of the common type of vertebrates, and the first concern of the *morphologist* should be to distinguish them carefully from the organs that are simply analogous and have forms and relationships that are of interest mainly to the *physiologist*.

Instead of comparing different animal species, one can follow Vicq d'Azyr who, after Galen, was the first to compare different organs of the same animal. His study clearly showed that the resemblances between the parts of our bodies are even closer than those of our arms and legs. Oken's school was especially concerned with the search for these resemblances, and since they do indeed exist, the principle of repetition provides a way for naturalists to obtain useful results. The limbs and vertebrae are the parts of the skeleton in which a similar repetition is seen most clearly, and this same repetition causes the organs that resemble one another to be disposed in series. They must also bear the same name, and the new kind of homology that this entails is what Owen called *serial homology* or *homotype*.

An understanding of homotype organs greatly simplifies the search for a common plan in the skeleton's structure. It is then seen that the multiple parts group themselves in similar segments, and if one has a good understanding of one of these segments it is possible to find rules governing the way in which all the others are constituted. For this reason, Owen assigned great importance to the determination of essential parts that make up the *vertebrate segment* to which he attached all the other parts of the skeleton and enabled him to arrive at a new way to enumerate the vertebrae in the cranium. This also permitted him to eliminate from the number of vertebrate parts a certain number of other parts that had accidentally been included in the composition of the internal skeleton. Some of these parts, as Carus had already shown, are dependencies of the skin, while others are part of the distinctive protective cover of certain viscera.

These comparisons were only the prelude to work that was necessary in order to arrive at what is considered an archetype. No single living being fulfills the role of archetype completely. Among the countless variations in the forms of parts, such as their differing positions, abnormal growth or reduction in size, or their atrophy or melding with other parts, it is necessary to distinguish what was accidental from what was essential. Only essential elements should be incorporated into the archetype, because this permits one to include all the forms under a common law without having to use special criteria for each individual group.

Once the archetype is established, one need only examine the types in question to identify the parts that definitively characterize that particular archetype, and if two types are compared, it becomes apparent that, after identifying their homologous parts, one need only relate them to the corresponding parts in the archetype. Thus, one can conceive of two kinds of homologs: those shared by the organs of idealized beings are called *specialized homologs;* those relating real organs to idealized ones that are modifications of the archetype are called *general homologs*.

Thus the fins of a porpoise are *special homologs* of the pectoral fins of fish and the wings of birds, but when these members are considered 'divergent appendages of the rib-derivatives of the archetype' their relations are said to be those of *general homologs*.

One can imagine an archetype for each of the main branches of the animal kingdom. Already in 1820, as we noted earlier, Audouin had tried to use a method analogous to the one that Owen later used to determine the general type from which all the arthropods could be traced. Audouin's results in the case of the external skeleton of arthropods and those obtained by Owen for the internal skeleton of vertebrates could be stated in the same basic terms: the skeletons are divided into segments that are fundamentally identical; they have the same division of the segments into central parts and appendages, the same repetition of these segments into linear series, and the same tendency to group themselves into more or less distinct regions. The juncture of these two archetypes confirms some of Geoffroy's ideas and shows the degree to which they conform to reality. Is was natural for Dugès to try, as Geoffroy did, to identify resemblances between the ideal vertebrates and articulates and in that way arrive at a more refined theoretical type of which the vertebrate and articulate would be only modifications. Nothing prevented them from applying to the archetype the methods of comparison and abstraction that Goethe, Audouin, and Dugès employed in order to arrive at more and more general organic types until they reached a point where they had eliminated all the relevant points of resemblances. The basic idea behind the attempts of Geoffroy and Dugès to determine what one could call an archetype of the animal kingdom is fully justified from a theoretical point of view by the obvious success of the studies of Audouin and Owen.

But what could be the significance of these archetypes to which they seem to have attached so much importance? We can get a clear insight into this by examining the methods that were used to determine them. Given that all the vertebrates and arthropods have certain general similarities, all the similar parts of these animals are compared one by one, and by examining all their modifications more and more closely, one can determine the extreme limits of these modifications. One then takes a kind of mean between the extremes and uses it to represent the archetype. A mean of this kind will be apparent whenever one addresses a zoological group that is relatively isolated from the others, as are several of the higher groups of animals. This archetype will become closer to the real forms as one addresses more limited groups. In this way, it will be easy to establish an archetype of the mammals, the birds, reptiles, amphibians, and boney fish, and deduce from a comparison of these forms an archetype of the vertebrates. But when one looks at the cartilaginous fish, the archetype of the skeleton is notably unrepresentative, and in the end, one must admit that all its distinctive elements have disappeared. One finds this to be true, for example, if one goes back as far as the *Amphioxus* or even the lampreys, even though they clearly have the qualities of vertebrates. If it is necessary to ignore most parts of an archetype in order to make it apply, it becomes apparent that the basis one has chosen in order to establish such a relationship is not broad enough. It is not entirely realistic, and even if it makes it easier to coordinate a certain number of facts, it is not adequate to tie them together in a meaningful way.

Let us go back to the basic facts and recognize that, as the amphibians seem to indicate, the skeleton of the vertebrates consists primarily of a dorsal cord to which various bones have been joined over a long succession of generations. Once the skeleton has reached a state in which it is able to undergo external modifications it does so without changing the number and essential relationships of its parts. It should be possible to deduce all these forms from a certain archetype that will include all forms except those that developed prior to the stage we are addressing. If one neglects these earlier forms, as one tends to do because they are of a lower order, one concludes that the bones in these groups of vertebrates are perfectly stable. This is the conclusion that Owen reached. He failed to recognize that this stability is only apparent because he unconsciously adopted a convention that automatically excluded anything that was not in accord with his assumptions. Unfortunately, the result was repeated appeals to the high-minded principles that led him to discover the archetype of the vertebrates:

'The unity of design implies that there was also a unity in the intelligence that conceived it. To ignore or deny this truth would cast an ever-lasting veil over human philosophy.

'The disciples of Democritus and Epicurus reasoned thus: if the world had been made by a superior mind or a pre-existing intelligence, that is to say a God, there must have been an *idea* and a *design* for the universe before it was created, and this in turn implies a *pre-conceived plan* for the order of nature even before the existence of things.

'From this, the followers of these ancient philosophers... having discovered no indication of an ideal archetype in any of its parts, concluded that there could have been no knowledge or intelligence before the beginning of the world. Today, however, the recognition of an ideal basic design for the organization of the vertebrates proves that the concept of a creature such as man existed before man made his appearance; the divine intelligence, in forming the archetype, foresaw all the modifications that it would undergo.

'The idea of the archetype is manifested in the organisms that are adapted to the varied nature of the surface of our planet, and this came about long before the existence of the animal species in which we see this design manifested today.

'Under what natural laws or secondary causes did the succession of species come to align itself? That is the question to which we have not yet found the answer. But if we can conceive of the existence of causes such as those of an all-powerful deity and see them personified in nature, the past history of the earth tells us that it has advanced by slow and majestic steps guided by the light of the archetype in the milieu of the ruins of former worlds since the time when the idea of vertebrates manifested itself under its old fish-like remains until the moment when it showed itself in the glorious form of the human body.'[2]

Even if we overlook the fundamental errors that, as we have already seen, have discredited the basic concept of the archetype, we cannot ignore all the dangers inherent in using such an argument. There actually exist, by Owen's own admission, several archetypes in the realm of animals. One could conclude with equal logic that each of these is the manifestation of a separate deity, and if one wants to consider each of these archetypes as only a particular thought of a unique creator, one finds it astonishing that such a small number of thoughts were responsible for populating the earth. But there are also groups that have no defined archetype whatever, unless this role is assigned to the simplest forms from which these groups evolved. What, for example, are the archetypes of the sponge, the coelacanth, or the worm? What we see in these different types is the gradual passage from a simple shapeless mass of viscous material to complex forms composed of parts that are disposed according to a rigorously determined order in creatures that are clearly constructed according to a common plan. One can follow the progress of phenomena that, step by step, have gradually gone through an infinite number of vague and indecisive forms to arrive at these well-defined, apparently immutable types.

---

[2] R. Owen, *Principes d'ostologie comparée*, 1855, p. 11. [The publication from which this quotation was taken was a French translation of the original, and the wording differs slightly from English versions on the same topic. Trans. note]

Leaving aside these primitive forms, one can deduce an archetype for the remaining creatures just as one can for the vertebrates or articulates. In this, however, one has a clear demonstration that the so-called archetype is not a *basic principle* achieved in a single step and carried on to infinitely varied forms, but the *result* of a long, slow evolution from simple primitive forms. One can no longer consider the homologies of various orders according to the primitive laws so clearly expressed by Richard Owen. On the contrary, these homologies still pose problems for which the naturalist must search for solutions.

We have just seen that Owen, like Geoffroy Saint-Hilaire, assumed that animal species can vary. According to Geoffroy, this variation resulted from the all-powerful action of external conditions, whereas Owen declared that we are still in complete ignorance about this problem; his concept of archetypes introduced an even more profound difference between his ideas and Geoffroy's. Geoffroy assumed that there was only a single plan of composition for the animal kingdom and that all living forms could be derived from a unique primitive form. As soon as one assumes several independent archetypes this becomes impossible. The variability of a particular species is limited to the modifications that are possible for its particular archetype. This limited variability is the compromise that one hoped would be found between the indefinite variability of the living forms, as Lamarck, Geoffroy, and Oken assumed but other naturalists considered excessive, and the absolute fixity on which the disciples of Cuvier insisted despite abundant evidence to the contrary, including numerous examples from paleontology. At the same time that Richard Owen was allowing for this limited variability Isidore Geoffroy Saint-Hilaire was developing a theory based on brilliant erudition and strict logic. Geoffroy's work was very impartial. It displayed a keen desire to find the truth, but when he wrote his *Histoire naturelle générale des règnes organiques* (1854–1862), he tempered this desire with a caution that now seems to have been a bit excessive.

Before Richard Owen tried to establish the archetype of the vertebrates and even before the word archetype had been coined, others had attempted to deal with the articulated animals much in the way that Richard Owen did. Obviously inspired by Geoffroy Saint-Hilaire's ideas about a unified plan of composition, Savigny, his companion on the trip to Egypt, had shown that the mouths of all orders of insects were made up of a constant number of identical the parts. In 1820, Audouin interpreted the crustaceans in terms of the theory of metamorphosis that Wolf and Goethe had derived from their studies of plants, and he enunciated two propositions, which at that time were very important:

'1. The different annular segments in the bodies of articulated animals are always composed of the same parts.

'2. All the differences we see in a series of articulated animals result from a similar or dissimilar growth of the segments, a union or division of the pieces that compose them, or from a maximum development of some and a rudimentary state of the others.'

This work demonstrated a unified plan for the composition of these articulated animals that was identical to that which Geoffroy Saint-Hilaire had proposed for the vertebrates, and at the same time it showed that the bodies of the former resulted from a repetition of parts that were fundamentally equivalent to one another. The archetype he proposed for them was in accord with Owen's principles, but in this case it was revealed with unusual clarity and invited us to study it more deeply.

The crustaceans possess a great number of members, all of which have highly varied forms and functions. In the crayfish, for example, there is a pair of peduncles that bear the eyes, two pairs of antenna, a pair of mandibles, two pairs of jaws, three pairs of claws, five pairs of walking feet, and six pairs of abdominal feet of which the last are transformed into flattened fins. Audouin succeeded in proving that all these parts are constructed in the same way, that they can be traced back to a typical form in the same manner as the segments of the body, so that the peduncles of the eyes, the antenna, the mandibles and the claws can be thought of as modified feet, a conclusion that Latreille immediately extended to the antenna and the chewing apparatus of insects. Audouin referred to all these varied parts as *appendages,* and their fundamental identity, which had already been demonstrated by comparative anatomy, was soon established by embryology through the important research of Rathke[3]. Through his studies of the crayfish, Rathke showed that all the appendages of that animal have the same basic form when they first appear and occupy the same position relative to the rest of the segment to which they belong. Little by little, they assume their definitive form and take on their specific functions. The peduncles of the eyes and the antenna are formed on the lower face of the corresponding segment, and only after it becomes an adult do they occupy a place above the mouth in a way that tends to mask their true origin. All the appendages do not appear simultaneously; the ocular peduncles, the two pairs of antenna, and the mandibles, that is to say the first appendages to the head are formed first and are followed successively by the others. In the same way, the head and the last ring of the abdomen were the first to appear, and all the others come out between these two. The last ones always appear between the last and next to last segments of the body. Rathke also pointed out another important fact: the parts that are formed first in the crayfish are the same that formed first in vertebrates; these parts occupy the future ventral face of crayfish and the dorsal surface of vertebrates. Thus, embryology confirms the hypothesis of Geoffroy Saint-Hilaire and Ampère that the vertebrates differ from the articulated animals, because they have taken on exactly opposite positions with respect to the ground.

The studies of Jurine, Thompson, and Nordmann, along with those of Henri Milne-Edwards, continued to add new support for these important discoveries. These capable observers showed that a number of crustaceans, especially the lower groups, underwent profound metamorphoses after emerging from the egg. While most of the more advanced crustaceans, like the crayfish, have all their segments

---

[3] Rathke, *Ueber die Bildung und Entwicklung des Flusskrebses*, 1828, in-folio, Leipzig.

when they hatch and do not undergo anything more than modifications in the form of these rings and appendages; others have to produce new segments before becoming adults. Milne-Edwards noted that, in this latter case, the various regions of the body, such as the head, thorax, and abdomen, can be equally incomplete and grow, each in its own way, along with the complete animal, and by adding new rings to its posterior.[4] Whatever its ultimate form may be, the young crustacean is born with only three pairs of feet that serve as a temporary means for swimming, but these represent the three first pairs of cephalic appendages, so that *the antenna and the mandibles (as with the jaws and legs) are destined to be feet for locomotion at a certain stage in the animal's existence.* One can say quite simply that these are modified feet.

In 1834, Milne-Edwards examined, compared, and interpreted all these modifications of form, metamorphoses, and differences in mode of development in a few lines that show how, even at that time, this illustrious scientist already had a profound sense of the relationships uniting living forms and did much to stimulate research and set the direction of zoological thinking in France.

'First of all,' he said, 'these diverse modifications do not appear to depend on any constant tendency inherent in the organism, and one could reasonably conclude that the development of each of these animals follows different laws, but that cannot be so because on studying these changes more closely, one sees that they can all be classed in a logical manner and, despite their great diversity, can be related to a small number of regulating principles that are also revealed in the kinds of metamorphoses that, as we have just seen, are displayed in the embryos of these animals.

'The changes that the young crustaceans undergo after their emergence from the egg can be thought of as complementary to the metamorphosis of the embryo. Although almost all of this metamorphosis takes place before the young creature has left the egg, it may be born somewhat prematurely and continue to change after its birth, so that it still arrives at structural changes that are analogous to those that it would have undergone during the embryonic stage of its life.

'These modifications are of two kinds: some consist of the appearance of one or more rings of their bodies and the members that stem from them, while other changes are manifested in the forms and proportions of parts that already existed before birth and either persist throughout its entire life or disappear more or less completely. 'All the decapods appear to be born with all their segments and appendages in place.[5] This is also true of certain amphipods [Edriophthalmes], such as *Amphithoe* and *Phronyme*, but other animals of the same group, such as the cymothoids and anilocra, emerge from the egg with only six pairs of legs instead of seven. In the case of the entomostracans, the young are much less advanced in

---

[4] H. Milne-Edwards, *Mémoire sur les changements de forme que les Crustacés éprouvent pendant leur jeune âge. (Annales des sciences Naturelle)*, vol. XXX: 182.

[5] It is now known that the Penæus, like the lower crustaceans, are born with only three pairs of appendages.

their development. In general, *one can distinguish only the cephalic members, and in this sense, they resemble the embryo of the crayfish near the beginning of its second period of incubation.* The thoracic and abdominal segments, as well as the limbs that are attached to them, appear only in succession, and the animals arrive at their final state only after having molted several times.'[6]

And further on:

'Modifications of the parts that already exist in the crustaceans at the time of their birth differ from one species to another, but they always tend to make each animal more and more distinct within the group to which it belongs, so that each takes on its own individual character. These animals resemble one another much more at the moment of birth than they do in the adult state, and, as a general rule, they become increasingly distinctive as they undergo more modifications during the early stages of their life.'

This offered an almost complete theory for the metamorphosis of the crustaceans. Even after fifty years, it requires no alteration other than to place more emphasis on some of the propositions that it contains. One could, for example, formulate the principles of these propositions in the following manner:

'All the crustaceans revert initially, either in the egg or after hatching, to a common larval form known as the *Nauplius*. There are now only three pairs of limbs that become cephalic appendages, mainly the antennae and mandibles.

'The nauplius represents only the head or a part of the head of the adult crustacean. The other segments of the body appear one by one in the posterior parts.

'These segments can be formed either in the egg or only after hatching.

'And finally, in each of the main groups, almost all the species go through a certain number of shared forms and their metamorphoses become increasingly complex as the adult form becomes more separated from the normal forms of its group.'

The doctrine of descent has since provided support for the laws deduced from these observations. When he stated them in their original form, Milne-Edwards was especially impressed by what he saw as a clear confirmation of the principle that the crustaceans are constructed on a single plan and are not the result of gradual complications by successive additions of new parts to a primitive nauplius which later became modified in diverse ways. What he visualized was essentially that the crustaceans are formed from an invariable number of segments. 'In principle, one can assume,' he said, 'that the normal number of segments making up the body of the crustaceans is twenty-one.'[7] All these segments can be traced back to a single ideal type of which they are only modifications. It follows that all the forms that succeed one another during metamorphosis are equivalent to one another and are always an accurate representation of the crustacean with the twenty-one segments that they tend to produce. Forms with fewer segments are merely anomalies; the nauplius and all the intermediate stages that separate it from

---

[6] H. Milne-Edwards *Histoire naturelle des crustacés* vol. 1, p. 197, 1834.

[7] *Histoire naturelle des crustacés,* vol. 1, p. 14, 1834.

the adult form are essentially transitory, and a crustacean that stops at one of these stages deviates from the basic pattern. In a word, the crustacean with twenty-one segments, in his view, is a fixed unit and each segment is only a fraction. Today, however,[8] it seems that the basic unit is probably the segment or zoonite, and the various varieties of crustaceans can have any number of component parts. In the hypothesis of a unified plan that Milne-Edwards adopted in 1834, a crustacean that hatches before having achieved twenty one segments is a crustacean 'that was born prematurely.' The hypothesis of descent does not preclude a crustacean from having any number of segments. Hatching should normally take place after the nauplius is formed, (and one could even visualize it being more premature). The segments would then be formed one by one after hatching. If it were otherwise, hatching would have been *retarded*, while the developmental phenomena leading to the crustacean of twenty-one segments would have been *accelerated*.

Such subtleties are no doubt open to question, but they are an excellent example of how easy it is to keep a scientific idea up to date by making minor alterations that keep it consistent with the increasing factual knowledge that a general interpretation must be able to accommodate. The theory that Milne-Edwards proposed in 1834 leads one to accept the theory of archetypes, and, even if the embryological phenomena observed in crustaceans can be expressed by a small number of rules, they do not violate his general principles. On the contrary, by accepting these subsidiary interpretations the highly varied phenomena involved in the development of the crustaceans is simply explained, as are those seen in the higher animals, by a simple acceleration of phenomena that differ in no way from those of reproduction by budding.

In 1845, Milne-Edwards added an important corollary to his interpretation of the crustaceans that implicitly reduced the significance of the number of parts and assigned greater importance to the different regions of the body. Following de Quatrefages' discoveries of the reproduction by division of the tiny marine annelid, the *Syllis*, and the remarkable budding of other annelids, the *polychaetes*, he showed[9] that the laws governing the growth of annelids are basically the same as those of crustaceans. He emphasized that the segments of both groups are formed successively and that it is always the next to last segment of the body or the last of each region that gives birth to the new segments. He continued:

'When the development becomes most active, as in the case of multiplication by budding of the *Syllis* and other polychaetes, one even sees a segment give rise directly to two or more zoonites that, by reproducing themselves in the ordinary way, constitute one or more intercalated series. The assemblage of segmented products then represents a series of groups of zoonites, each one of which extends by its posterior part as did the unique series in the preceding case... This phenomenon which, in the class of annelids is manifested only during production

---

[8] See my work on *Les colonies animales*, p. 505, 1881.

[9] Milne-Edwards, *Observations sur le développment des annelids* (*Annales des sciences naturelle*, 3rd series vol. III, 1845, p. 174).

of new individuals by way of budding… is also seen in the development of the embryo… In the case of the crustaceans, for example, there appear to be three of these systems or series of genetic systems in which elongation can continue after formation of the first ring of the following series, and it should be noted that these three groups correspond precisely to the three large regions of the body of these animals, the head, the thorax, and the abdomen.'

A short time later, Milne-Edwards would show that certain regions of the bodies of various sedentary annelids function like similar parts of the bodies of crustaceans, but at the same time he established that the growing number of body segments of the annelids and their agamous reproduction are only two very similar manifestations of the same phenomenon and that various regions of the body of crustaceans correspond, in the case of the annelids, to the new individuals that split off to lead an independent life and can therefore be considered distinct individuals.

Like the crustaceans, the most diverse types of annelids resemble each other during the earliest periods of their development. This remarkable coincident in the progress of genetic phenomena of two very different types inspired Milne-Edwards to offer the following reflections:

'Zoological affinities are proportional to the duration of a certain parallelism in the progress of the genetic phenomena of various animals, and creatures that are still in the course of formation cease to resemble each other sooner if they belong to distinct groups at a more elevated level in our system of natural classification and that the dominant essential characteristics of each of these divisions would reside, not in some particular feature of the permanent organic form of adults but in the more or less prolonged existence of a common primitive constitution, at least in appearance.'[10]

This is far from the principles of Cuvier, who wanted to employ all the recognizable characteristics of the species in classifications. Embryology now took on an important role in classification. Creatures having the same larval forms were henceforth recognized as being related, and if this is still considered an idealized parental relationship, it is evident that there will be no need to change anything in the newly derived formula when it becomes necessary to recognize that parentage must be understood in the true sense of the word. Serres, in France, and the natural philosophers in Germany had made a similar proposal when they said: 'As they develop, all the higher animals pass through forms that are analogous to those that remain permanent in the lower animals.' This new formula is more complete and precise, and, as we know, progress in science almost always consists of replacing an idea that is only partially true with a more general rule that offers more satisfactory explanations. The formula of the natural philosophers assumes a unique type of development; that of Milne-Edwards includes not only the proposals of the German scientists but also that of Von Baer, who had identified several

---

[10] Milne-Edwards, *Considerations sur quelques principes relatifs à la classification naturelle des animaux.* (*Annales des Sciences Naturelles*, 3rd series vol. 1: 65. 1884.)

types of development. Milne-Edwards had an advantage over Von Baer in not having to place limits on the possible types of development and being able to consider embryonic characteristics at all levels of the classification, as has been attempted several times more recently.

Embryology had already rendered important services to taxonomy when it made it possible for Thompson to show that barnacles, which were assigned to the mollusks by Cuvier, to the annelids by Latreille, and to a special group of their own by Blainville, were in fact true crustaceans.[11] Nordmann had shown that the anchor worms, which were universally thought to be worms, also belonged to this same group of crustaceans.[12] The phases of development have often revealed unexpected relationships between creatures that are quite different in the adult state, and naturalists have put so much trust in clues of this kind that the danger is now that they may take apparent similarities in the larval forms as real identities.

In summary, despite these successive modifications of what we take a crustacean to be, Milne-Edwards' final, definitive theory can be stated as follows: All crustaceans are constructed according to the same plan. Their body is composed of the same number of segments that form identical parts. The various crustaceans differ from each other only in modifications of the form of these segments or the parts composing them. In general, these modifications do not appear in the individual until the embryo has reached a more or less advanced stage of development, so that most crustaceans, notably those that belong to the same group, begin by resembling one another and then differ more and more as their development advances. The segments of the body are formed successively, but they may form slowly or more or less rapidly, and eclosion may take place at any particular stage of this development. During this multiplication of segments, each part of the body acts as an independent organism.

These proposals could be extended to all articulated animals. There seems to be an archetype for the arthropods, just as there is for vertebrates, but it is different, and the existence'of several organic types claimed by Cuvier seems to be confirmed. As Milne-Edwards pointed out, however, most of these simple prototypes of the articulated animals result from a successful application to crustaceans of the method employed by Geoffroy Saint-Hilaire in his study of the vertbrates. 'The theory of analogs,' he said,[13] 'became famous through the work of its author, Geoffroy Saint-Hilaire, and through the new trends it instilled in comparative anatomy. As we see, it has reduced most of the difficulties that previously encumbered studies of the integumental skeleton of the crustaceans. And even though the utility of applying these basic theoretical principles to entomology was already demonstrated by the studies of Savigny, Audouin, etc., further proof was provided by the simplicity of the corollaries that sum up the causes of the innumerable differences seen in the integumental skeleton of the crustaceans.'

---

[11] Thompson, *Zoological research and illustrations, or natural history of nondescript or imperfectly known animals*, 1831.

[12] Nordmann, *Mikrographische Beitrage zur Naturgeschichte der wirbellosen Thieve*, 1832.

[13] *Histoire naturelle des crustacés*, vol. 1, p. 50.

The hypothesis of a unified plan of composition imposed limits on the extent of each of the branches of the animal kingdom and made it easier to attribute the resemblances one observes in various animals to a common cause. Would it not be possible to treat their innumerable differences in the same way? Already in 1827, Milne-Edwards had indicated the way to do this in the articles he wrote for the *Dictionaire classique d'histoire naturelle.* Not only did he formulate a law that has since taken on increasing importance, but he was the first to indicate in a precise way an unforeseen similarity between the laws of political economy and those of general physiology. This opened a very promising path that gained wide acceptance after the work of Charles Darwin showed that it can yield unexpected results. For Milne-Edwards, the cause of the diversity of animals is the *division of physical work* between their constituent elements. In Darwin's view, the origin of species must be sought in the *competition* driven by *population growth* and the survival of the best adapted individuals in the resulting process of natural selection. Economists now consider the division of labor through specialization the surest means of competing in the market place. Far from losing its value from the advent of Darwin's doctrine, one can say that Milne-Edwards' ideas gained added strength and durability. On the other hand, the division of labor assumes *association,* a principle which, as we have seen, Dugès applied to the animal kingdom in 1831. In my book on *Les colonies animals et la formation des organisms*, I have pointed out the importance, in terms of the evolution and gradual development of living forms, of the laws that have prevailed during the formation of organic types. It also helped explain embryonic phenomena and even what we call *individuality*. Thus the parallels are prolonged, and whenever the laws of political economy are applied to morphology, they seem to have fruitful results. It is evident that the aspect of the question that touches on the way in which the four major modes of distribution of the characteristic parts were realized, the four organic types described in *Les colonies animales* could not exist if one limits one's interpretation to the hypothesis that the initial forms of different organic types are subject to only minor modifications: that is the point of view taken by Dugès and Milne-Edwards. These naturalists had no doubt already made discoveries of their own and recognized a certain number of facts that permitted one to establish a theory for the way these organic types were formed. Nevertheless, they accepted, just as Cuvier, Geoffroy Saint-Hilaire, and Richard Owen did, the hypothesis that all organic beings are direct products of the Creator's will, and it is only to these *pre-established* types that they apply the theory of the division of physiological labor. They expressed this in terms such as these:

In 1827, Milne-Edwards said:[14] 'the bodies of certain animals always have identical characteristics, and, so far as I can see, they have no distinct organs... The fresh-water polyps share a structure of this kind.... The bodies of these animals can be compared to a workshop where each worker would be employed in carrying out similar work and where, as a consequence, each member will have an

---

[14] *Dictionnaire classique d'histoire naturalle*, vol. XII, article Organization, p. 339, August 1827.

influence on the whole, but not on the overall outcome. Moreover, experience has shown that in dividing one of these beings one does not change its behavior; each fragment continues to live as before and can form a new animal.... On the contrary, when life begins to manifest itself by more complex phenomena and the final result produced by the interplay of the different parts of the body is perfected, certain organs acquire a particular kind of structure and cease to act in the manner of the whole. The life of the individual, instead of being the sum of a more or less large number of elements of the same nature, results from the combination of acts that are essentially different and are performed by distinct organs. The diverse parts of the animal economy all pursue a common goal, but each does this in its own appropriate manner, and the more numerous and developed the faculties of the being the more diverse the structure and the farther the division of its labor is extended.'

Milne-Edwards laid out his thoughts more clearly when he later wrote:[15]

'The principle that nature follows in perfecting beings is the same as that which is so well developed by modern economists, and, in its works as well, just as in the products of human enterprise, one sees immense advantages in the division of labor.'

Milne-Edwards applied these principles of the division of labor to different systems of organs and primarily to the teguments.

'The external form of the body, like the internal part, has a series of modifications that we can explain by the same principle already discussed. As we have just said, it is initially similar to the rest of the parenchyma but soon acquires different properties and constitutes a distinct membrane the inner face of which attaches to all the active organs of locomotion and its surface becomes the seat of sensitivity, respiration, and several other functions.

'As the faculty of perceiving light has become more perfected in the higher classes of animals, it has also become more localized. The same is true of the sense of hearing and smelling, but the envelope encasing the body still serves as an organ for movement and feeling. It determines the form of the body and protects the internal organs from the harmful external agents. This division of labor is developed even farther in the most advanced animals and colonies where a specialized system is especially designed to protect the soft parts and provides the essential function of locomotion. The tegumental membrane, instead of serving diverse purposes, no longer acts as a tactile organ, but has only a small number of other functions, such as hindering the evaporation of liquids from the body.'

In this passage, the principle of the division of labor is not applied to independent individuals that play separate roles but to homogeneous elements that have no clear individuality and form heterogeneous parts suited to particular functions. There is no connection or relationship between forms in which there is little division of labor and those in which it is more advanced; the author was obviously not trying to establish any kind of genealogical relationship between the internal skeleton of the vertebrates and the external skeleton of the articulates. The principle

[15] *Histoire naturelle des crustacés*, vol. 1, p. 5, 1834.

of the division of labor is really only a way to explain the relationships of certain observed facts. It is a kind of metaphysical expression of a *process* by which forms of life that are initially simple become more complex.

This can be seen very clearly in the way Milne-Edwards made use of the process. A division of labor in response to external conditions causes individuals that start out identical and independent to become increasingly modified and grouped according to their different functions. Changes of this kind necessarily imply that these forms of life have undergone a gradual transformation. Moreover, his statement of this proposition, which was initially metaphorical, has gradually taken on a more concrete meaning. This is certainly true in the explanation Milne-Edwards proposed for the development of the nervous system. He stated that:[16] 'In studying the extensive series of articulated animals, one sees that the parts by means of which these beings perceive sensory signals have gone through the same series of modifications to which we referred in discussing the tegumental apparatus and organs of metabolism. The initial form of the nervous system is a cord extending through the whole length of the body. At that stage, all the parts act in a uniform manner and, when the animal is divided into several parts, each continues to feel and to move just as it did when the body was whole. A further step in the division of labor leads to the localization of the faculty of perceiving sensory stimuli and of several other actions in definite parts of the system, the existence of which then becomes necessary to maintain the integrity of functions of the system as a whole. Finally, in the case of more highly perfected animals, each of the senses is supported by its own specialized medullary fibers. The faculty of producing willful bodily movements is concentrated in other fibers of the same system. The faculty that stimulates the action of these diverse parts is localized in corresponding parts of the nervous system, while the coordination of movements is exercised by other independent instruments. In a word, all parts of the sensorial apparatus contribute separately and in a different manner to controlling the action of the whole organism.'

Milne-Edwards took the same approach when dealing with teguments and their inferred morphological development. After studying the diverse modifications of the nervous system of the crustaceans, he summarized them according to the *law of centralization* that Serres used to represent the successive modifications that the nervous systems of insects undergo in the course of their development.[17]

'The nervous systems of all the crustaceans are composed of medullary centers the number of which is equal to that of the limbs. All the observed modifications of nervous systems, whether they be from differing periods of incubations or through development in different species, depend mainly on the degree to which these centers have been integrated. The agglomerations of nervous tissues are connected from the sides inward toward the median line, as well as in a lengthwise direction, but their development may also be arrested in a certain number of its centers.'

---

[16] *Ibid.*, p. 126.

[17] *Ibid.*, p. 147.

The agreement between facts revealed by comparative anatomy and those furnished by the embryological development of a given individual already implies that the various states of the nervous system could have developed from a primitive state in which all the ganglia were identical. This is certainly the idea that emerges when, after considering the tissues or organs, Milne-Edwards concluded by saying the following about the bodily segments themselves:[18] 'According to what I have said at the beginning of this chapter about the process nature follows in perfecting the being, one can expect to find at the lower extremity of the groups of animals considered here, species in which all the segments making up the body would be identical, not only in form and structure but also in their functions, but would then differ more and more as each one becomes adapted to a particular use. This is what one notices on comparing various crustaceans, but these animals offer no examples of such extreme uniformity and complexity.'

A division of labor can govern the development of all the segments, just as it does the organs and tissues. It must then be followed by a kind of morphological adaptation resulting from more or less extensive modifications in the form of the segments. But for Milne-Edwards, these segments are not, as they were for Dugès, distinct individuals. They are simply parts of the body, and they occur in numbers that are always inherent to the crustacean. Despite the segmentation of their bodies, the crustaceans are as indivisible as vertebrates. Many naturalists still think of the arthropods in this way, and from the point of view of transformism, we have seen that this idea serves to surmount the problem of having to explain the origin of each organic archetype as the result of a special act of creation.

We have already noted that in studies of the articulated animals, just as in those of vertebrates, Geoffroy Saint-Hilaire's method has been used to define in a stricter, more rigorous, and complete manner Cuvier's major branches of life and to determine the limits of modification of which they are capable. This same method is used to determine the laws governing these modifications. Until now, the principle of connections has been applied mainly to the solid parts and makes it possible to trace their disposition to a common type. It can be applied equally well to the internal organs and other soft parts of the body.

Cuvier had made the nervous system the basis for a methodical classification of animals, while Emile Blanchard wanted to determine all possible modifications of the nervous system within a given branch and to indicate the relative importance of these characteristics as guides to classification. He demonstrated that in insects it is based on a constant type and generally tends to become more or less concentrated during metamorphosis. This concentration follows definite laws, so that one finds 'remarkably regular family characteristics in the degree of centralization of the medullary nuclei.'[19]

---

[18] *Ibid.*, p. 20.

[19] Emile Blanchard, *Recherches anatomiques et zoologiques sur le systeme nerveux des animaux sans vertebras Annales des sciences naturelles,* 3rd series, vol. 5, 1846.

His studies of the connections of the nervous system brought him remarkable insights into the relationships of organs. He showed, for example, that the antennae that seem to be absent in arachnids are actually represented in these creatures by the small claws of scorpions and venomous stingers of spiders. These are the only appendages that have nerves connected to the brain in the same way as the antenna of insects and crustaceans.

After Blanchard completed Savigny's studies of the mouths of dipterous insects, de Lacaze-Duthiers undertook a study of the complex appendices in their posterior extremities and showed that the female genitals are constructed according to a unique plan identical to that of the mouth.[20] The multiple parts that compose it are the direct result of the development and morphological modifications of the solid parts of a zoonite. Thus, in the case of the adult arthropods, and especially the more highly developed species, numerous studies have made it possible to trace their very diverse organizations back to a single basic plan. Throughout the entire class of insects, the number of segments of the body remains constant. The same is true of the number of regions of the body and appendices devoted to a specific function. In the arachnids, the total number of segments of the body is less constant. It is especially variable in the case of the myriapods, even though the head has a constant composition. Finally, if there is a certain fixity in the higher crustaceans, one notes in them a great variability in the constitution of various regions and the number of appendages that serve analogous functions. On the other hand, the segments of the body do not always develop simultaneously, and that in itself is enough to cast doubt on the proposed immutability of the type and suggests that, if this immutability really exists in certain groups, it has been acquired and must still be considered a *result* rather than an *inherent cause.*

The excellent studies of the annelid worms by Savigny, Audouin, Milne-Edwards, and de Quatrefages already show that these animals have uniformly organized segments, the numbers of which can vary widely, so that one can hardly see that they bear any similarity to an archetype. It is quite a different matter when one descends from the annelid worms to those in which the segmentary structure is less distinct. Blanchard's patient and skillful studies of intestinal worms, as well as those of Quatrefages on the *Planarius*, show that the essential traits that Cuvier attributed to the segmented animal have faded away and disappeared. Nevertheless, the idea of types is so tenacious that one does everything possible in order to make these animals conform to a rule from which they always manage to escape.

Cuvier defined the mollusk branch less rigorously than he did the vertebrates and arthropods. Milne-Edwards' studies of these animals revealed an unexpected imperfection in the construction of their circulatory systems; and studies of the nervous system of bivalves by Duvernoy and Blanchard and of gastropods by Quatrefages and Lacaze-Duthiers point to a well-defined molluscan type in which,

---

[20] Lacaze-Duthiers, *Recherches sur l'armure génitale femelle des insects Annales des Science Naturelles*, 3rd series, vol. XII to XIX, 1829 and following years.

as Lacaze-Duthiers showed, the connections [homologies] among the parts are as well defined as they are in the other groups. Unfortunately, instead of limiting this now well-defined type to the cephalopods, gastropods, razor clams, and bivalves that are true mollusks, attempts have been made to include, (as had been done with the worms), anything that presents more or less vague analogies [similarities]. In this way, one strives passionately to find the characteristic traits of mollusks in brachiopods, tunicates [sea squirts], and bryozoa without realizing that, in attempting to transform the criteria so that other organisms can be included, the type loses its identity. If the theoretical type is defined on the basis of the hypothesis of fixity of species and one attempts to relate different species without making allowances for possible changes of their structures, it is impossible to explain forms that embryology and comparative anatomy have clearly shown can be related only as the result of evolutionary changes.

The difficulties of Cuvier's theory of major branches were already pointed out in 1822 by Blainville who, while assuming the absolute fixity of species, considered animals as members of a certain number of *types* with gradations between them on a scale comparable to the one proposed by Bonnet. He assumed that in each of these cases the organizations of the species could have undergone successive degradations that rendered their characteristics unrecognizable because there were no distinguishable breaks between the degraded and higher forms of each type. Faith in the genius of Cuvier, however, enabled certain minds to overcome these difficulties. One of the master's most eminent disciples, Louis Agassiz, became a strong advocate of the theoretical concept of types, and the moment arrived when it was necessary to show how a very resolute partisan of the absolute fixity of living forms could turn it into a doctrine of zoological philosophy.

Chapter XVI

Louis Agassiz

Philosophical consequences of the hypothesis of fixity of species. – The possibility of setting up a classification that demonstrates the existence of God. – The existence of a plan of creation is inconsistent with the doctrine of transformism. – Arguments in favor of the fixity of species. – Weaknesses of these arguments. – Characteristics of the graduated zoological divisions. – A new definition of species. – Inconsistencies between this definition and observed facts. – The reality of species. – Causes of the physiological isolation of species.

Louis Agassiz[1] applied an idea analogous to that of Owen's archetypes to all aspects of the so-called natural system. He proposed that each of our species, each of our genera, each family, and each biological type represents a distinct concept of the Creator, and, as a result, all these groups of individuals have an equal reality. Classification, far from being what Lamarck considered an 'art form' that varied according to the views of the individual artist, is an immutable edifice reflecting the will of the Creator. This was also the opinion of Cuvier and the naturalists who, like him, were making the study of natural processes the supreme goal of science. The various zoological groups 'have been arranged according to categories laid out by a divine intelligence.'[2] Richard Owen rejected the idea of final causes and deduced that the existence of archetypes of the vertebrates in itself offered sufficient proof of the existence of God. Louis Agassiz took a similar approach. The existence of a series of plans that served as models for living beings demanded an intelligence capable of conceiving these plans. 'The rational, intelligible links between phenomena are a direct proof of the existence of a God capable of thinking, as surely as man manifests his ability to think when he recognizes these natural relationships.'[3] Just as it is our intelligence that enables us to penetrate this order of nature that Agassiz attributed to the existence of God, it is the existence of our own intelligence that offers proof of the existence of God. The Swiss scientist was not far from saying: 'I think; therefore, God exists.'

---

[1] Louis Agassiz, *Contributions to the natural history of the United States*, 1857; *Essay on classification.* London; 1859; *De l'espèce et de la classification en zoologies*, Paris, 1862.

[2] L. Agassiz *De l'espèce et de la classification en zoology*, p. 8.

[3] *Ibid.*, p. 14.

A. McBirney, S. Cook and G. Retallack, *The Philosophy of Zoology Before Darwin: A translated and annotated version of the French text by Edmond Perrier*,

Louis Agassiz assumed a pre-established harmony between our intelligence and the universe: 'The human mind is a unification of nature, and many things that seem to be the result of our intellectual efforts are only the natural expression of this pre-established harmony.'[4] This was his basis for a natural classification: 'The systems that we have set up under the names of the scientists who proposed them are really only translations of the Creator's thoughts into the language of humans. If this is so, the faculty of human intelligence that enables it to adapt instinctively to the conditions created for it also enables it to interpret the thoughts of God. Is this not conclusive proof of our affinity with the divine spirit? Must not this spiritual and intellectual relationship with the all-powerful make us think profoundly? If there is any truth in the belief that man is made in the image of God, nothing is more appropriate for the philosopher than to study the operation of his own mind and endeavor to bring his works in closer accord with divine reason. By studying his own intelligence, he can reach a better understanding of a higher intelligence of which his is only a feeble branch! At first glance, this suggestion may seem disrespectful. But who is truly humble, he who delves into the secrets of creation, classifies them according to a scheme he proudly calls *his* scientific system, or one who arrives at the same end but claims an affinity with the Creator and, with profound gratitude for this sublime gift, endeavors to interpret the divine will with which it is his privilege to be in communion?'[5]

This passage is of great interest. It is an unusually complete expression of a philosophy of nature that can be traced, first from Linnaeus to Cuvier, and then from Cuvier to de Blainville and Agassiz without ever being as clearly formulated as it is here. Unlike Schelling, Agassiz did not assume that this identity of the human spirit with that of God allowed him to say: 'To philosophize on nature is to create one's own nature.' Far from rejecting factual evidence, as the German philosopher had done, he studied objective observations and concluded from their relationships that we possess an intelligence that is in accord with God's will and enables us to see and understand the divine origin of all these relationships. It is not just studies of objective facts that were being neglected but also those of natural forces and their effects on living beings. We no longer had to search for the causes that have brought life to its present state; there was only one cause, God, and He acts without intermediaries. We no longer have to look for the origins of those organic features that our scalpel reveals to us: 'there are organs that have been preserved only to maintain a certain uniformity in the fundamental structure… They are not there to serve any particular function but to maintain a certain uniformity of plan. It reminds one of the way architects design our edifices with classic exterior styles, harmonious proportions, and perfect symmetry, but with no practical purpose.'[6] Thus, there is no ultimate cause in the universe, only a single purpose: the fulfill-

---

[4] *Ibid.*, p. 9.

[5] L. Agassiz, *De l'espèce et de la classification*, p. 8.

[6] *Ibid.*, p. 12.

ment of the Creator's will. The role of the naturalist is simply to reassemble the facts that express this will and to coordinate them in systems that are our way of expressing God's will. Louis Agassiz boldly expresses here a doctrine that, more than once, has been the underlying cause of the hostilities aroused by sincere and quite legitimate attempts to arrive at some sort of understanding of the origin of life and the laws governing its evolution. Moreover, he sought to set limits on these attempts: 'If it is some day proven that man has not invented but has only described an existing systematic ordering of nature and that the relationships and harmonious proportions in all parts of the organic world *have their guiding spirit in the mind of the Creator*, or that this plan of creation that defies our most solemn wisdom is not the product of inescapable physical laws but, on the contrary has been freely conceived by an all-powerful intelligence and was thoughtfully worked out before being manifested in tangible external forms, and finally, if it is demonstrated that premeditation preceded the act of creation, we shall finally, once and for all, be able to ignore all the distressing theories that force us to call upon the material laws of nature in order to explain the great number of marvels we see in the universe and, by banishing God, leave in His place the monotonous action of constant physical forces that adjust everything to an inevitable destiny.'[7] There must be something in this *inevitable destiny,* this *fatalism,* that seems to imply descent with modification [transformisme] and that many minds find frightening. One defends God's liberty by thinking that in this way one is safeguarding one's own. All the philosophical arguments, all the aspirations of the heart and mind, are powerless to change what we are or the relationships that can unite us with the world or with God. And, in the end, what does it matter whether we have reached our present state of perfection in one way or another? Is anything to be gained by willfully deceiving ourselves in this regard? On the contrary, is it not wiser to try to use all the means at our disposal to penetrate the secrets of our origin and the laws governing our progressive development. Only in that way can we gain a clearer knowledge of the justification we can reasonably claim for our existence and of the destiny that human society aspires to. It can tell us how best to achieve that destiny and contribute to the evolution of our species. Is this not the way to arrive at an intimate knowledge of the collective entity we call humanity and find a rigorous definition, independent of all faiths, of the rights and obligations of all individuals making up the human community? Is this not the way to inculcate the definitive morality that the bewildered human mind, plagued by errors, prejudice, and violence, has always tried to establish? As it slowly evolves, it is beginning to emerge from the shadows of our ignorance.

Agassiz's mind was too scientific to deny the premise that physical forces are capable of creating or modifying living beings. He needed a way to reject their determining role and wanted to make the evidence as convincing as possible. His arguments can be summarized as follows:

---

[7] *Ibid.*, p. 10.

1. Today, we find all kinds of animals living in identical conditions. To infer that they owe their character to these ecological conditions is to assume that the same cause can produce widely differing effects.
2. The same types of animals can be found living under highly varied conditions; this shows that the way these creatures are organized is independent of physical agents.
3. From one pole to the other along all the meridians, the bodies of mammals, birds, reptiles, and fish share a single structural plan. Equally marvelous plans are found in the arthropods, molluscs, radiata, and many types of plants. This infinite variety of creatures with bodies constructed on a single unified plan could not be the result of blind forces that lack intelligence, are incapable of thought and cannot conceive of space or time.
4. All animals are manifestly the result of four basic plans that are linked to one another by the fact that they all began with an egg in which they developed independently of external forces. Even though they were initially identical, they are now very diverse.
5. Members of the same genus, family, or class are found in very different climates, and despite their great variety, analogous relationships exist between the animals of various lands, even though there is now no genealogical relation between the species of the same genus, the genera of the same family, the families of a single class, or the classes of a single major branch. It is not logical to attribute the characteristics that members of a given order share in common to the effects of physical forces that are supposed to produce the same type of form regardless of where they now reside.
6. The four major branches of the animal kingdom appeared simultaneously with their distinctive characteristics despite the fact that they were all living under identical conditions. Thus, from the very beginning there were clearly distinguishable classes, families, genera, and species within each of the four branches.
7. It is difficult to see gradations between the major branches or classes, but one can recognize gradations within a given class, and they are in accord with the timing of their appearance in the geological record. 'There again, a new and startling proof of ordering and gradation was already established at their origins and has been preserved through the ages in various degrees of complexity reflecting the structure of the living creatures.'[8]
8. Some species, genera, orders, and even closely associated neighbors, can be living under very different conditions. Some live in a variety of ecological settings, while others appear to have a more restricted geographical distribution that cannot be attributed to the effects of local conditions.
9. Regions with very similar climates can have identical fauna and flora, or they may have very different ones that have always flourished in these localities. This is clearly inconsistent with the idea that the animals and plants could

[8] L. Agassiz, *De l'espèce et de la classification*, p. 43.

have first developed from couples with accidental variations that were propagated only in that locality. In other cases, the fauna and flora in settings that differ little from those of another region are found to include very specialized types. The marsupials of Australia are a good example. These differences cannot result from the effects of the environment, because all the members of the fauna and flora should have been modified in the same way.

10. The different types within a given series of forms are often found in such widely separated regions or in such different paleontological orders that one can hardly infer that they are linked by any genetic relationship. Moreover, the capricious compositions of these series implies that there has been a conscious choice of combinations rather than a continual action of blind forces. The fact that the members of these groups are disseminated over the entire surface of the globe indicates that the intelligent being that created them prevailed in all places and all times.
11. Despite the diversity of the conditions to which the species are exposed, the species of a single family tend to have a rather uniform size. This indicates that ecological conditions were not an important factor in determining their size.
12. Of the many known species, the only ones that have developed significant variations did so only through the will of a powerful intelligence, man. This demonstrates that another powerful intelligence has intervened to modify the fauna and flora.
13. The mental functions of animals are basically similar to those of humans, and it follows that all are the expression of a spiritual principle that cannot be attributed to physical forces; instead, they testify to the existence of a universal intelligence.
14. This intelligence is clearly manifested in the precisely regulated relationships between individuals of the same species, between the various animal species and ambient conditions, and between animals and vegetable species inhabiting the same niche, as they do, for example between parasites and their hosts.
15. The various embryological phenomena, such as metamorphosis and asexual reproduction that we shall examine later, clearly show that physical-chemical forces have had only a minute effect on the development of the individual.
16. There is a remarkable relationship between the organic types that succeed each other in the paleontological series: certain *synthetic types* combine within themselves characteristics that are later found only in different, widely separated types. Others, the *prophetic types,* have organs that, in an imperfect form, seem to forecast the appearance of new types having more fully developed organs and functions that until then had been absent. For example, the pterodactyls, the flying reptiles, seem to have prophesized the birds that arrived a short while later but have no direct relationship to reptiles. Others, the embryonic types, have permanent characteristics that will be only transitory in their successors. The fossils of similar types found in more ancient strata show that such combinations had already been envisaged long before a wise and prescient intelligence created them as living forms.

17. There is a parallelism between the order of succession of animals and plants through geological time and the gradations seen in the organized life of today. One can recognize in this a will that has overseen all development of nature from beginning to end and allowed a slow, gradational process to prepare for the introduction of humans as the crowning creation in the animal world. There is a similar parallelism between the order of appearance of the animals over time and the phases of embryonic development of species living today. The same concept has directed the development of both.

Louis Agassiz concluded that:

'Far from owing their origin to the continued action of physical influences, all living beings have made their appearance on Earth by virtue of the *direct* action of the Creator.

'The products of what are generally called physical agents are *the same throughout the world,* and *they have remained the same* through long periods of geological time. Organized life, by contrast, has always *differed both temporally and spatially.* There can be no causal link or relationship between two types of phenomena with such different characteristics.

'The combination in time and space of all these profound concepts is not only a clear manifestation of intelligence, but it also proves premeditation, power, wisdom, grandeur, prescience, omniscience, and providence. In a word, all the facts and their natural relationships proclaim the only God that man could know, worship, and love. Natural history will one day become the analysis of the Creator's thoughts as they are manifested in the animal and plant kingdoms just as they have been in the inorganic world.'[9]

Richard Owen believed that the archetype was a direct product of divine thought but that secondary modifications due to the effects of the environment had been able to modify it in a thousand ways. Agassiz believed, as we have seen, that it is possible for this divine intervention to appear even in the simplest of phenomena. This is a direct consequence of the hypothesis of the fixity of species. No one else has developed this conclusion so thoroughly, and no naturalist has assembled a greater number of arguments to support it, but we must ask whether the arguments he presented necessarily have the significance that he attributes to them. Not a single one of the phenomena that Agassiz invoked has received a natural explanation. The mixtures of different animals living in seemingly identical conditions, the persistence of similar forms under highly varied living conditions, the superposition of genetic characteristics on the secondary characteristics of families, genera, and species – all these are direct consequences of Lamarck's law of heredity. In a recent work,[10] I have related the formation of the principal organic types to recognizable causes, and shown that these types must have appeared and developed simultaneously: the constant mixing of different organic forms one observes

[9] L. Agassiz, *De l'espèce et de la classification*, p. 218.

[10] Perrier, 1881, *Les colonies animals*, p. 714.

in all the geological epochs is a consequence of this elementary fact. All the observed facts about geographical distribution have proved to be arguments in favor of the theory of evolution. Just as Agassiz has shown, the various relationships between individual animal species, the external world and the living creatures with which they find themselves in contact are simple adaptive devices that are essential consequences of natural selection. Today, there is general agreement that no species can remain absolutely immutable when it is exposed to strong modifying effects and that the variations of domesticated animals are no different from those of wild animals; instinct and intelligence can explain one just as well as the other. The parallel between paleontological and embryonic evolution has become one of the most fruitful lines of evidence for evolutionary theory. In a word, all this scholarly discussion is turned to the profit of the doctrine of evolution that it claimed to refute: it seems clear that in our times the creative acts have acted only through the intermediary of conflicting properties inherent in the living substance and the conditions in which each organized individual is called upon to live; nothing indicates that it could ever have acted otherwise. Our new concept of the organized world, even though it recognizes no primary causes, in no way diminishes the majesty of the marvelous organization of our universe. Whether one seeks to understand the motives of the Creator or the processes by which He puts them into effect, both goals are equally worthy of the human intellect.

However that may be, even if we concede that the various divisions of the animal kingdom could be the result of a divine plan in which each division corresponds to a special creative category, this hypothesis would imply that each unit must have its own particular role to play in the overall plan. For this reason, Agassiz was attempting to identify the characteristics of each branch, class, order, family, genus, and species in the entire animal kingdom.

He found that the characteristics of the major branches of life conform to the same simple factors on which the abstract *plan of organization* was originally based. The manner in which the divisions were set up or, one might say, the materials composing them, became the defining characteristics of the classes and were based primarily on their anatomical structure. A plan carried out using this same material added a somewhat greater complexity to its meaning. One must search through a wide range of complications to find the characteristics of orders that are totally gradational between one and another. The general modifications that the external form can undergo without changing the general structural plan become a characteristic of the family. They include not only the general modifications of the external form, but also the modified forms of parts of the body. These modifications determine the characteristics of the genera, and all that remains is to define the species.

In taking this approach, Agassiz separated himself completely from the naturalists who based the idea of species on the ability of the individuals of the same species to give birth to new offsprings that are just as productive as their parents.

'As long as we have no proof,' he said,[11] 'that all our varieties of dogs, all our domesticated animals and cultivated plants are derived from a unique species, pure and uncontaminated, there will be doubt about the common origin and unique line of descent of all our human races. It would be illogical to assume that sexual procreation, even when the product is fertile, is unquestionable evidence of specific identity.

'In order to justify this assertion, I shall ask whether any unprejudiced naturalist would now dare to state:

'1. That it has been proven that all the domesticated varieties of sheep, pigs, oxen, lamas, horses, dogs, fowl, etc. are derived from common stems.
'2. That to say that these varieties are the result of a mixture of several primitive species is an unacceptable hypothesis.
'3. That varieties, such as the Shanghai chickens and our common European chickens that have been imported from distant lands and never had any earlier contact with one another are now incapable of interbreeding.

'Where is the physiologist who can seriously affirm that the limits of fertility between distinct species are known sufficiently well to make it the distinguishing mark of their specific identity? Who could say that the distinctive characteristics of the fertile hybrids and those of pure-bred species are so evident that one could trace the primitive traits of all our domesticated animals or even those of all our cultivated plants?'

Agassiz is obviously on dangerous ground in making this argument for the fixity of species. If primitive species can interbreed to the point of being able to produce what we call our domesticated species, even when human intelligence has been their sole author, it must follow that these species can vary, but one can get around this difficulty by saying that we are wrong in thinking that each of our varieties of dogs, cattle, or pigeons forms a single species and that we can therefore discount the fact that they are able to interbreed. Agassiz was saying, in effect, that God created species in the same way he created genera, families, and other categories of beings between which the naturalist sees resemblances – that there is no genetic link between the individuals of the same genus, family, or order, and that there is no necessity for a genetic link between individuals of the same species. The first individuals from which they descended were created separately in great numbers. At the moment when these independent individuals were created, the species were as limited as they are today. One must therefore look for the distinctive signs of the species in the structural and morphological features of the individuals and not in their ability to reproduce, which is a simple consequence of the amount of resemblance that one sees between individuals.

Louis Agassiz was pushing the logical consequences of his system to the very limit. In accepting the fixity of species as a fact, he made the concept of species completely hypothetical and entirely dependant on his interpretation of creation.

[11] L. Agassiz, *De l'espèce et de la classifications*, p. 262.

He saw in this the idea that individuals of a given species living in a particular geological period had similar relationships to one another and to the surrounding world. And yet, the parts of their bodies had the same shapes, proportions, and ornamentation, which indicates that, when submitted to the same influences, they all vary in the same way. Thus, his definition of a species required a detailed knowledge of the organization and mode of life of the beings it comprises.

Agassiz could have simplified this definition by accepting Linnaeus' hypothesis: 'We count as many species as came two by two from the hands of the Creator.' But that would have required identifying species according to a range of features that extend beyond those used in the conventional system. It would have required that there be a genetic relationship, a true consanguinity, between all animals of the same species, even though there are no such relationships between animals that belong to the same genus but have been created independently of one another. This would have disrupted the harmony of the system, and the reasoning would lead an advocate of the fixity of species to make a choice that Cuvier declined to make when he said: 'The species is an assemblage of individuals that share the same parents and resemble those parents as much as they resemble one another.'

By denying the possibility of mutability, the hypothesis gave the zoological classifications an apparent precision that many found attractive, but nature, in its constant mobility, tends to emphasize all parts of the links by which we try to join them. Louis Agassiz could define the systematic divisions of the various degrees only by giving his definitions a flexibility that rendered them illusory when they were applied to factual observations or used in making comparisons that were difficult to justify: every definition of species suffers from this same general submersion of facts under a theory that tries to make them conform to one's preferred way of interpreting observational evidence.

The fact is that there exist groups of individuals that can be mixed together in all proportions, and one cannot draw a precise line of demarcation between forms that still maintain their individual identity. It is equally true that all combinations of individuals that differ in some way are always sterile. Thus, there is a definite demarcation between individuals of the first group and those of another. Each group that is isolated in this way constitutes a *species*, but one finds all intermediate stages of these unions between absolute fertility and the complete infertility of the unions. Individual members of the same species generally present an almost complete identity of structure while varying somewhat in their relative size, proportions, color, and behavior, so that they may sometimes differ from one another more than they appear to differ from individuals of another species. Most of these differences can be attributed to external circumstances, while the fundamental resemblances have nothing to do with the effects of their surroundings. Differences between individuals of the same species can be transmitted to their descendents, so that all the progeny reproduced within their group or with others they resemble always present the same assemblage of permanent, inherent characteristics that distinguishes them as varieties of their species. They form *races* that can be almost as fixed as species when only individuals of the same race are interbred, but they can be altered by mixing with individuals of a different race. It is a fact that individual members of an animal species have different degrees of resemblance to one another that cannot be explained by the *present* effects of ambient conditions, resemblances on which all the structure of our zoological divisions is based.

There is no problem, of course, if this effect goes no farther than the individual in which it is first reflected. It would be necessary to conclude that the world can be explained only by appealing to supernatural causes. But the effect of the surroundings does not end in that way. A certain number of the modifications it has engendered are transmitted to the progeny where they become increasingly stabilized as one generation follows another under conditions that favor their preservation. With time, they become fixed in future generations, and the individuals in which they have acquired a degree of stability can then be placed in the most varied conditions without losing their characteristics. Here again, we are faced with facts that undermine several of the arguments that Agassiz invoked in favor of his system. Once this is seen, the problems must be posed in a different manner.[12]

The fertility of a union depends simply on what the sperm of the male can normally contribute to the egg of a fertile female. Only the results of this process are known; we know nothing about how they are accomplished and what conditions are required for them to be effective. We do know, however, that a very slight modification in the conditions in which the egg finds itself placed suffices to prevent its fertilization. The female body can be modified in many ways without reducing its capacity for fertilization of the egg. Some kinds of modification, however, promptly lead to an inability to hatch eggs. This might be a good way to investigate how different races within the same species can drift apart and continue to bear resemblances to one another even though they are no longer capable of interbreeding. In this way, new species could result from the same causes as new races. They differ from the ordinary races only in that modifications of the latter preserve some parts of the body while modifications of a new species will result from biological conditions that permit the sperm to fertilize the egg. These conditions can probably be determined, and the problem of identifying them is within the scope of normal experimental physiology.

If species are constituted in this way, all resemblances between the different species are explained as the result of inherited characteristics. Their permanence results from the tendency of procreation to offset individual differences, so that the resemblances become increasingly stable with each successive generation. Natural selection explains the relative isolation of species, as well as their narrow adaptations to external conditions. Knowing this, we can arrive at a better understanding of both the apparent fixity of specific forms and their variability over

---

[12] This paragraph reveals the depth of Perrier's understanding of the intrinsic genetic factors that govern organic development. Imperfect as it may be, his explanation is far better than Agassiz's. It also shows how the debate was being focused more on the mechanisms of inheritance. (Trans. note)

time. The main problem this leaves us with is that of determining the past conditions that could have produced and locked in specific characteristics.

When one carefully examines the lines of evidence on which zoologists have based their reasoning until now, it is apparent that they have drawn almost exclusively on studies of relatively developed, distinct types of animal life, such as the vertebrates, arthropods, and molluscs that are the basis of modern zoological philosophy. Our knowledge of these animals has reached such a high level of perfection that it should be possible to sort them according to general propositions similar to those used in the physical sciences. The long-neglected study of simpler animals, almost all of which are lumped together in the major branches of animal phyla [zoophytes] or categories laid out by Cuvier, tends to enlarge the scope of our scientific perspective and shows that questions that are thought to have been long-since resolved have scarcely been broached and are open to renewed consideration. It is essential that we seize this opportunity to go back and start again at the beginning.

Chapter XVII

The Lower Animals

> Progressive discoveries relative to the lower animals – Trembley: the fresh-water hydra. – Peyssonnel: the corals. – Cuvier: the pennatule. – Lesueur: the siphonophores. – De Chaisso “ the alternating generation of the salpas. – Sars: the alternating generation of the hydromedusas. – Steenstrup: a theory of alternating generations. - Van Beneden: digenesis. – Leuckart: polymorphism. – Owen: parthenogenesis and metagenesis. – M. de Quatrefages: geneogenesis. – Milne-Edwards’ theory of reproduction. – General theory of the phenomena of asexual reproduction.

Naturalists have long recognized that most invertebrates fall into a small number of basic groups. Aristotle, as we have seen, distinguished various types that he logically grouped together and even observed the behavior and metamorphosis of several insects. Almost everything that was known about them during the middle ages was based on his writings, but there was little understanding of its meaning. The metamorphosis of insects prepared the mind to accept bizarre interpretations that had no basis in fact. When one has seen a butterfly born from a caterpillar, it is not difficult to believe, as Aristotle did, that caterpillars are born from green leaves or that worms are formed in the mud that they inhabit – the same mud from which, according to Genesis, God brought forth man.

It was impossible to sort out meaningful ideas and separate them from the complex history of the lower forms of life at a time when there were no objective observations or instruments to augment the power of human senses. It was not until the XVIIth century that the use of magnifying lenses enabled Malpighi, Swammerdam, and Leuwenhoek to study the fine structure of a body and recognize things that, until then were shielded from view by their small size. Malpighi concerned himself mainly with anatomy and embryology and Swammerdam with the metamorphoses of insects. Leuwenhoek used a lens to examine a great variety of objects, and he was the first to point out the existence of Infusoria and to study an animal that budded like a plant, the fresh-water hydra. Trembley’s later research would make it famous. At the same time one of his students at Hamm discovered spermatozoa. The initial impact of all three of these discoveries had a considerable repercussive effect.

A. McBirney, S. Cook and G. Retallack, *The Philosophy of Zoology Before Darwin: A translated and annotated version of the French text by Edmond Perrier*,

The Infusoria have since been the subject of endless speculation. It was hoped that one might find in them material that was in the process of evolving itself. They were found to be living atoms. They prolonged the debate about spontaneous generation, but in the end, they helped us to identify the material that makes up the living elements [cells] of our bodies.

The fresh-water hydra has been the first example of arborescent organisms of which coral is the archetype. This led to a better understanding of what these remarkable organisms might be.

Spermatozoa which at one time were believed to be a rudimentary form of animal life, provided arguments in support of the doctrine of preformation as long as development of animals by epigenesis had not been rigorously demonstrated. They have become the basis for all our thinking about conditions that are essential for the development of life.

But these observers did not immediately recognize the fundamental importance of such factors. One could not fully appreciate the role of spermatozoa without having first noted how the female produced an egg and how embryonic material developed. It was not until 1824 that Prévost and Dumas first identified the division of the egg cell, and soon after, in 1827, von Baer discovered the egg of mammals. The fresh-water hydra had been almost completely forgotten until 1744, the year in which Trembley's memorable study was published. The infusoria, however, continued to be the subject of vague speculation until Ehrenberg took up the earlier work by Frederic Müller and, in 1829, offered one of the principal arguments against the hypothesis of primitive protoplasm. It was he who dealt the first blow to the widely held belief that a swarm of insects, worms, molluscs, and even certain mammals such as rats, could arise spontaneously from inert substances. Rédi showed that worms in meat hatch from eggs laid by flies, but he could not surmount the difficulties he encountered in his studies of parasitic worms. He believed that they were formed at the expense of the sensitive soul of their host. He thought that Harvey's famous axiom: '*Omne vivum ex ovo*,' [all life comes from an egg] was misleading. Most naturalists continued to believe that worms were produced by spontaneous generation. Many even wondered whether Adam's parasites were created at the same time as he was, and it was not very long ago that medicine finally consented to accept round worms and tape worms as forms of animal life like other worms.

Rédi's studies were among the first of a series that succeeded in drawing a line between organic and inorganic life, between a living substance and inert material. Although this delineation became sharper over time, this has certainly not been true of that between plants and animals. There were a few notable works that only tended to obscure their differences.

During the XVIIIth century, Linnaeus' aphorism: '*Minerals grow, plants grow and live, animals grow, live, and have senses*' was considered the final word. However, certain creatures of the sea were an embarrassment for the ancients; coral was one of the chief ones. Theophrastus, Dioscorides, and Pliny did not hesitate to make it a plant, but Orpheus believed that we should assign it a more noble origin. It was hardened and colored by the blood of the Gorgon Medusa, and Ovid recounted that, soft and flexible in water, it became hard only when exposed to air. Boccone showed in 1674 the weakness of this view, but he made coral out to be stone. Ferrante Imperato (1699) and Tournefort (1700) placed it back among the plants, and their opinion seemed to prevail definitively after 1706 when the Count of Marsigli reported that he saw branches of coral flourishing in fresh water. A third opinion appeared in 1713 when Rumphius, in his *Amboinische Raritätkämmer*, spoke of polyps that resemble plants. This view was formally expressed in 1723 by a young doctor from Marseille, Peyssonnel, a friend of Marsigli, who saw the supposed 'flowers' [polyps] of coral eating and moving about and compared them to the marine anemones that are so common along our coasts. But Réaumur gave this new opinion a very cool reception. For him, coral was a plant that produced an internal shell, in the same way that the snail produces an external one. To him, the outer part of the coral was alive; its stony axis was a dead concretion. Réaumur could not imagine that a branching concretion such as coral could have anything but a vegetative origin (1727). The ability to put out buds and branches and to be divided without dying were, in his time, the essential characteristics of plants, but the definition of plants was destined to undergo an abrupt change.

In 1740, Trembley came across Leuwenhoek's fresh-water polyp and was very intrigued by this strange creature. Thinking that he was the first to observe it, he decided to determine its nature. His first impression, based on the green color of the first individuals he observed and on their root-like tentacles, was that they were plants, but when he saw that these plants could move and eat food, he began to have doubts. It seemed to him that in order to resolve the problem he had only to find out whether the polyps are capable of budding and could be reproduced from cuttings. He then undertook a beautiful series of experiments in which hydra cut into pieces returned to a form identical to that from which they were cut and continued to live and reproduce their missing parts. He observed that his polyps, by successive budding, could form a group of about twenty individuals; that on being split longitudinally each strip became a new polyp. In this way the initial polyp acquired several heads and mouths, just like the hydra in the fable. From that it gained its scientific name, Trembly's *polyp with horn-like arms*.

All these experiments established that the hydra shared with plants the power to bud and to reproduce from branches, but a creature that can move, capture and devour its prey, and change places at will cannot belong to the plant kingdom; it is an animal. Thus, there can be branched animals just as there are branched plants. The coral must be an animal of this kind, and it seems that Peyssonnel was right. Réaumur, Bernard de Jussieu, and Guettard quickly seized the occasion this offered to study marine polyps, and in the end, Peyssonnel's views won the approval of the Paris Academy of Sciences. It was recognized that coral, bryozoans [flustres], and other 'marine stems' are aggregates of animals that are born by budding one upon the other to form a structured society. It was still difficult, however, to accept the idea that Linnaeus, in the twelfth edition of his *Systema naturæ* (1766), tried to offer as a compromise: for him, the corals and bryozoans are plants that vegetate under water but produce flowers [polyps] that are animals. This was one of his last attempts to deal with the observational evidence. It first had to be shown that the factual evidence had not been compromised, because Gaspard Wolf, who undertook his studies of embryology (1759) in order to find out if there might not be something comparable in the development of the animal to what was observed in plants, did not give a thought to polyps. It was the same with Goethe, who could not have failed to see in the animal communities that would soon be referred to as *colonies*, the exact repetition of this type of plant [archetypes] that he had so proudly proposed.

Trembley's studies lent support to the analogous research of his kinsman, Bonnet (1741), but the work of the latter was based on quite different forms of life, the *Tubifex*, fresh-water worms that are very similar to earthworms but have a simpler organization. Like hydra, if a *Tubifex* is cut into pieces, each piece can complete itself and become another worm. A single *Tubifex* could be cut into eight successive parts and repair itself so quickly that, according to Bonnet, one could obtain 2,985,984 new worms from a single one in six months. In one case, this skilled experimenter is even said to have succeeded it making a head grow where the creature's tail was originally located and a tail at the opposite end, so that it was turned end over end. These studies clearly confirmed the animal nature of hydra, which were found to have features analogous to those found in polyps. Later, Gruithuisen and Otto-Frederic Müller stated that another worm similar to the *Tubifex*, the *Nais*, could separate spontaneously into several individuals, the original individual being able to divide itself near the middle into two parts or to produce a chain of new individuals from its posterior part. In 1788, Otto-Frederic Müller added another bit of information to his earlier observations: he discovered a marine animal, the *Nereis prolifera*, later named the *Autolytus prolifera*, that, like *Nais*, spontaneously splits into two, but in this curious species, as well as in another that Müller had overlooked, the two individuals resulting from this parting differed from one another.

In 1828 and 1830, Dugès[1] observed phenomena in the lower worms *Planaria* that were even more similar to those that Trembley noted in hydra: he saw that in some species an individual could divide itself transversely into several others that remained united for a more or less prolonged time in a kind of annulated form, but in this worm, the rings soon separate from one another, as they do in the hydras, and then go on to live separately. This observation must have contributed to the ideas that the naturalist from Montpellier developed in his *Mémoire sur la conformité organique*.

[1] *Recherches sur l'organisation et les mœurs des Planaires Annales.* (*Annales des sciences naturelles*, 1st series, vol. XV, 1828), and *Aperçu de quelques observations nouvelles sur les Planaires et plusieurs genres voisins* (*Annales des sciences naturelles*, 1st series, vol. XXI, 1850).

The mode of development and reciprocal relationships of closely associated creatures like the corals revealed many astonishing insights to the naturalist who, until that time, had studied only the higher animals.

In 1803, Cuvier was studying a strange organism, the sea pen,[2] that emerges from a large living plume that thrusts its stem into mud on the seafloor and spreads bristles into the water in the form of large discs. He had recognized that these discs supported numerous polyps similar to those of the corals, but he also observed that all the polyps making up a sea pen are controlled by a single will and that they accomplish all the functions of nutrition in common. He concluded that the sea pen should be considered a composite animal. He extended this same reasoning to all the colonies of polyps, each one of which he considered a composite animal or, even better, an animal with several mouths and a single body.

In 1813, the illustrious traveler, Lesueur, published in the *Journal de Physique* an account of some of the remarkable animals he had collected during his long voyages. He drew attention to gelatinous organisms of varied and complicated forms that are known siphonophores.[3] He considered them to be floating colonies of medusas, an interpretation that was later adopted by Lamarck and Blainville.

In 1819, Adelbert von Chamisso, an ardent traveler, imaginative novelist, outstanding poet, and pains-taking naturalist, had pointed out a completely unexpected phenomenon in the reproduction of *Salpa*. This unusual animal of the class of tunicates, is as transparent as the water in which it lives like a gelatinous penguin equipped with variously placed appendages. It is able to swim by expanding and contracting its body. The different species of salpas belong to two general types, one of which is able to reach the size of one's fist and live a solitary life, while the other much smaller variety forms elegant rings or long, commonly phosphorescent chains. These chains had already attracted interest, because all the individuals that form parts of them coordinate their swimming movements with such precision that the entire chain gives the illusion of an animal directed by a single will. The aggregated salpas associated in chains are sharply distinguished from solitary salpas by both their external appearance and certain features of their organization. Chamisso announced to naturalists that, despite all these differences in form, size, and behavior, the aggregated salpas were daughters of the solitary salpas that they in turn reproduce, so that in the case of the single animals, the daughters never resemble their mother but rather their grandmother, and that the successive individuals produce in turn a single offspring or a multitude of twin offsprings destined to live together united by their members. One would almost take this to be the invention of a novelist's imagination, and even von Baer, accus-

---

[2] The sea pen is classified as follows: Phylum cnidaria (Coelenterata); Class Anthozoa; Subclass Alcyonia (soft coral); Order Pennatulacea (sea pen). 'Pen' refers to the resemblance of the vane to the plume of a quill pen. (Trans. note)

[3] Siphonophores are classified as follows: Phylum Cnidaria (Coelentrata); Class Hydrozoa; Order Siphonophorae. They are the Portuguese man-of-war and its relatives. (Trans. note)

tomed as he was to bizarre transformations of embryos, did not dare to sanction the traveler's account.

The questions posed by the observations of Cuvier, Lesueur, and Chamisso soon took on a broader importance and were so interwoven with one another that they called for a common response. In 1828, Michael Sars, who later became a pastor at Kinn and Mauger in Norway, discovered a kind of polyp with the external form of a hydra. He gave it the name 'scyphistom'. At the same time, he described another polyp, the 'strobile', that differed from the first in that its cylindrical body is divided into a series of superimposed rings each of which resembled a small medusa. Some years later, in 1835, he recognized that as *Scyphistom* grew it was transformed into *Strobile*, and that each of the rings of the *Strobile* metamorphosed little by little until it took the form of a small jellyfish that eventually detached itself from the rings that supported it and swam freely in the sea. Sars gave these small medusas the name *Ephyres* and followed their external transformation until, in 1837, he finally obtained the larger medusas known as Aurelies and Cyanées. Cuvier had placed the polyps and medusa in two very distinct classes within his branch of radiata, but these two classes were later merged into a single unit. One immediately notes that we are dealing here with a series of phenomena that are quite analogous to but even stranger than those that Chamisso had observed. It was now a matter of finding an explanation or at least a rule that governed their development, and several naturalists hastened to do this. Professor Lovén in Stockholm soon discovered that the arborescent colonies of other aquatic polyps, the *Campanularea* and the *Syncoryne*, also produce medusas that grow on them like flowers on a plant and then become detached.[4] Von Siebold, Dujardin, de Quatrefages, Desor, Van Beneden, and Max Schultze made similar observations that considerably extended and coordinated the outstanding work of Allman. The fact that the animals of a given form can give birth to animals with completely different forms was henceforth firmly established.

It will be recalled that the history of the round worms [helminthes] or parasitic worms has many unusual and still largely unexplained features. Swammerdamm,[5] Bojanus,[6] von Baer,[7] and Carus[8] saw that lower worms in the form of tadpoles, *Cercaria,* or even the familiar round worms [helminthes], *Distoma*, are formed in the interior of other living organisms that are also parasites. Fröhlich,[9] Zeder,[10] and von Siebold noted that a ciliated embryo quite different from its parents comes

---

[4] S. Lovén, *Observations sur le développement et les metamorphoses des genres Campanulaires et Syncoryne*. (*Annales des sciences naturelles,* 2nd Series, vol. XIV, 1841).

[5] *Biblia naturæ*, p. 75, fig. 7 and 8 of plate 9, 1752.

[6] *Isis*, Bd. I, 1818, p. 729.

[7] *Nova acta Academiæ Leopoldinæ*, vol. V, p. 2, 1826.

[8] *Nova acta Academiæ Leopoldinæ*, vol. IX, p. 75, 1835.

[9] *Naturforscher Stuck*, 25, p. 72.

[10] *Göze's Naturgeschichte der Eingeweidervürmen*, Suppl., 1800.

from the egg of the *Monostoma* and *Amphistoma*, and, according to Siebold's observations, the developing embryo already contained an organism with a quite distinctive form.

In their study of the class of solitary worms known as *Cestodes*, Pallas and Göse found astonishing resemblances between the *Cysticercus*, a short worm equipped with a large vesicle at one extremity, and the true tapeworm. In 1762 Bonnet[11] showed that the tapeworms should not remain indefinitely in the same host. One wondered whether the long-standing mystery of the reproduction of these creatures does not present phenomena similar to those found in the salpas and aquatic polyps. The time had come to coordinate all these marvelous facts. The young scientist, Japetus Steenstrup, who at that time was a reader at the academy of Sorö and has since become a professor at the University of Copenhagen, accomplished this task in 1842 and made a special effort to find a single law that would relate the nature of the reproduction of the salpa, the medusa, the tapeworms, and trematodes.[12]

The dominant feature in the reproduction of all these creatures is that a sexual organism of a given form gives birth to asexual creatures that do not resemble their parent but reproduce among themselves by a kind of budding or by division of their bodies to form new sexual beings like their parents. In this way, the sexual and asexual forms alternate regularly from one to another. Steenstrup called this *alternation of generations*, and he determined in a quite ingenious way the significance of the different forms that followed one another.

Sars and Lovén had seen in the scyphistoma a true polyp with a vastly simpler structure than that of the medusa. In their opinion, the polyp was a larva of which the medusa was the perfect form. They believed that the medusa, like insects, reached their definitive form only after having undergone one or more metamorphoses. Their metamorphosis resembled that of insects that pass through several forms and, like the medusa, go through two or more successive generations. Steenstrup showed that the scyphistome and the medusa are equivalent beings, one asexual, the other sexual. The sexual form produces eggs but dies once it has ensured the birth and growth of the larva. Parental care of this kind is confined to the asexual scyphistoma, which is nothing but an older member of the generation whose development it is destined to ensure. It is a creature condemned to celibacy in the interest of the brothers to which it is fully devoted. Steenstrup called it a wet-nurse. At the same time, in the case of bees, ants, and termites, only a certain number of the eggs laid by the females produce sexual individuals; the others produce neuter workers charged with raising the young and doing all the work that ensures the community's existence. Among these insects the workers are

---

[11] *Considérations sur les corps organizes.* (*Œeuvres d'histoire naturelle et de philosophie,* éd. Fauche, 1779, vol. III, p. 37).

[12] *Ueber den Generationsvechsel, oder die Fortpflanzung und Entwickelung durch abwechselneden Generationen, eine eigenthumlehre Form der Brutpflege in den niederen Thierclassen*, Copenhagen, 1842.

distinguished from sexual individuals just as the latter are distinguished from one another. Thus, it is not surprising that scyphistoma, the neuter medusa, differs from its mother, *Aurelia*, which is sexual. The same reasoning can be applied to the *Distoma* and even more so to salpa. It seems, therefore, that the singular phenomena of alternating generations comes back to the general rule that they are probably due to simple differences in the form of the individuals, differences that are analogous to sexual differences and to a way of raising the young for which the insects are examples.

Steenstrup's theory, which was based on evidence that he and his predecessors observed, enjoyed a very favorable reception. It has since been questioned, modified, and developed, but there can now be no question that it is in full accord with the results of a number of recent studies. Among the salpa, the aggregates develop from the eggs of solitary individuals. In the case of the aphids, Balbiani affirmed that the formation and fecundity of the eggs precedes the appearance of the individual that seems to have engendered them. In 1881, my studies[13] of the conditions of reproduction in certain types of colonies led me to state that the egg is an intrinsic property of these aggregations and not that of a particular individual. Various observations, notably those of Weissmann, Varennes, and the still-unpublished work of Rouzaud, have recently led to the recognition that in colonies of polyp hydra, the egg is produced in the parts of the colony that cannot be associated with any particular polyp, and it is only long after the appearance of the eggs that produce them that medusa succeed in maturing and becoming fecund. But, like all explanations based on the finality of phenomena, the theory of alternation of generations as it was developed by the illustrious Danish zoologist applies only to relatively rare cases in which an adaptation – an accord – has been established between two very general categories of phenomena that are otherwise unconnected: first, the formation of the egg inside the animal or in a colony and, second, the reproduction by budding of this animal or colony.

One finds that in a given zoological group there can be cases intermediate between budding that is produced in a completely independent manner and that in which it is linked to the formation of eggs. There are also intermediate cases in which all the individuals born of this budding are identical to their parents, as are many aquatic polyps and annelid worms, and those in which they are profoundly different. According to Van Beneden, the two modes of reproduction, from eggs and by budding, are a general phenomenon of which alternating generations are only a special case.[14] This highly capable professor working in Louvain coined the term *digenesis* for this general phenomenon.

Leuckart made an interesting application of Milne-Edwards' principle of division of physiological labor to this important concept of digenesis by adding to

---

[13] E. Perrier, *Les Colonies animals*, p. 726 and following.

[14] Van Beneden, *Mémoire sur les cestoïdes* (*Bulletin de l'Academie de Bruxelles*, 1847, p. 106.)

alternating generation the notion of *polymorphism*.[15] The individuals that produce eggs and those that produce only buds can have various roles to play in adapting to different living conditions. From then on, each one must take on an appearance and characteristics according to its function: alternating generation is only a special case of these varied processes of adaptation. Steenstrup had already decided that it is really a phenomenon of the same order as that which stems from the physiological differences between males and females, between sexual and neuter individuals in colonies of bees, ants, and termites, and even between the neuter members of these colonies when they have different roles to play. The dissimilar individuals born from one another do not necessarily separate: they can remain united and in this way form colonies in which the members have more or less diverse structures. It was in this way that Leuckart explained the astonishing organization of siphonophores, which are true mixtures of polyps and medusas but possess their own distinct individuality. The siphonophores, in turn, help explain the sea pens, which are colonies of coral-like polyps that Cuvier took to be animals with several mouths. The exceptional phenomenon that appears to have produced alternating generations is considerably elaborated: it can operate on the regular constitution of organisms so that various parts are merely individuals adapted to particular functions. It is in this way that a siphonophore can include various individuals that serve as feeders, prehensory organs, means of locomotion, and reproductive organs, all of which are polyps or medusa that have been modified for their special function by taking on a *form suited to their function*. Leuckart opened a fruitful way to approach the problem, and even though he did not really pursue it as far as he could have, a close link would be established between the theory of the constitution of the siphonophores and other composite animals as Leuckart understood it and the theory of the constitution of arthropods formulated by Audouin and Milne-Edwards. This link had already been established by Dugès even before the question of alternating generations had arisen: Leuckart's rule of polymorphism is essentially only an application to new or better observations of the principles that Dugès developed in his *Mémoire sur la conformité organique dans l'echelle animale* published twenty years earlier.

Even if it is shown that the animals possess two different modes of reproduction and that these modes result in organically dissimilar forms in the same animal species, this does not explain why the phenomena that depend on these two modes of reproduction are so frequently found in close association. Richard Owen, following a separate approach, wondered whether sexual and the asexual reproduction, which he called *metagenesis*, may not be related to one another. He tried to show this and explain the curious faculty of reproducing without previous fertilization that Leuwenhoek and Charles Bonnet had observed in a female plant louse. This type of reproduction without fertilization or, to use another of Owen's terms, *parthenogenesis*, has since been recognized in bees, yellow jackets, gall

---

[15] Leuckart, *Ueber den Polymorphismus der Individuen oder dir Ercheinungen der Arbeitstheilung in der Natur*. Giessen, 1854.

flies, several types of diptera and butterflies, as well as some of the crustaceans, rotifers, and several lower types of worms. It proved to be more wide-spread than was at first thought and became the basis for the illustrious English scholar's grand theory.[16]

Parthenogenesis, however, is only an illusion: according to Richard Owen, all ontogeny [évolution] is based on the union of a male cell and a female cell. After fertilization, the female cell – the egg – divides, and all living creatures are only assemblages of cells issuing from this division of the primary, original cell [l'élément primitif] repeated a great number of times. But this division of the constituent cells of the living being is nothing more than reproduction. It persists because each cell, on dividing, bequeaths to the cells that replace it a part of the activity that the egg has received from the fertilizing cell, the spermatozoid. The fertilized egg owes its activity entirely to the latter. However, the fertilizing power of the sperm is limited. It can provoke only a limited number of divisions and extends to only a finite number of anatomical cells. This accounts for the limitations of size and life span that one observes in all forms of life. In certain cases, all the anatomical cells born from the division of the egg are employed in the construction of a single individual. This is what happens in higher animals; in other cases, the fertilizing power of the sperm is not yet exhausted once the individual has been created. This individual is always a female. Males are produced only when the fertilizing power is on the point of reaching its limit. Up to that point, the reproductive power conserved by the females that follow one another can manifest itself in diverse ways. These females sometimes go on to produce eggs that can develop without new fertilization, as has been observed in fleas, bees, *Daphnia*, etc. In some cases they produce internal buds that organize themselves into new individuals, as we see in the trematodes. Sometimes they push out external buds that can detach themselves and become either separate, independent beings or remain united. In either case, the new individuals can take on special characteristics, according to their diverse functions. If they separate, one is faced with the phenomenon of alternating generations; if they remain united, they develop colonies such as those of the aquatic polyps, siphonophores, corals, bryozoans, composite sea-squirts, and tape worms.

The theory of parthenogenesis, as it is usually understood, has a very general connotation. It links a multitude of facts whose relationships had not even been suspected. The development of the individual, as it is seen in the higher animals, can be completely understood within the total assemblage of phenomena in which the formation of colonies, alternating generation, and parthenogenesis play equal parts. All the phenomena of reproduction can be related to a single basic type with modified details, and they are all based primarily on fertilization. Unfortunately, as Huxley, Carpenter, and Quatrefages have pointed out, this basic principle is now untenable. It is claimed that, under favorable circumstances, the faculty of reproduction without fertilization can be prolonged, if not indefinitely at least for a

---

[16] Richard Owen, *On parthenogenesis*, 1849.

very long time in the case of the female plant louse. The hydra have a similar faculty of budding. There is no way to attribute limited fertility to sperm, but we know that in a great number of lower creatures, possibly including the infusoria, reproduction is always accomplished without fertilization and this act, limited to the fusion of two protoplasms of identical appearance, is often confused with the phenomenon known as conjugation or the fusion of two gametes. This seriously undermines the basis of the theory of parthenogenesis, but it does not follow that none of the relationships that Owen identified exist. There are two notable features in the initial stages of development of most plants and animals: first, the division of the primary cell, the egg, into ever larger numbers of derived cells, and, second, its fertilization. Richard Owen recognized a causal relationship between these two generally associated phenomena, and he believed the governing one to be fertilization. But this was an arbitrary choice; the habitual coincidence of these two phenomena may very well be only a form of adaptation. Fertilization can become necessary for development under certain conditions without always having been indispensable to it, and from there on the important, dominant factor is the segmentation of the egg that is, in effect, the most general aspect of the process. This phenomenon leads back to a property that is common to all living cells [éléments vivants] that are capable of developing. They can separate as soon as their continued nutrition enables them to reach a critical size. This property suffices to explain all the phenomena and to join them together,[17] as the illustrious scholar – who has justly been called the English Cuvier – sought to do.

There again, a slight modification and minor rewording suffices to restore the value of a theory that seemed to be on the point of being lost, and as has been pointed out, later theories relative to the phenomena we are studying should be taken as special cases and important corollaries of a more general theory that they complement and render more important. It is quite true that the need of cells (and the organisms that they constitute) to divide into distinct individuals when they have achieved a certain stage of development, determines the existence of two modes of reproduction: one that requires fertilization and the other that does not. The phenomena of reproduction that are most general and that do not require fertilization can be designated by the name chosen by Richard Owen, *metagenesis*. When living species combine various degrees of these two modes of reproduction that can be independent, there is, as Van Beneden said, *digenesis*. If the individuals that form without previous fertilization are based on a cell more or less similar to an egg, there is *parthenogenesis* in the absolute sense of the word. When the various individuals produced from a fertilized egg take on different functions and when there is a division of labor among them that is necessary for the preservation of the species, they take on different forms and aspects of the essential physiological functions. As Leuckart proposed, *polymorphism* accomplishes its task of complication, and one particular case is called *alternation of generations*. It is equally true, as Steenstrup thought, that alternation of generations can have the effect of

[17] See my work *Les colonies animals et la formation des organisms*, p. 701.

producing asexually individuals that play the role of wet-nurse to those that are produced sexually and are really their siblings.

But metagenesis can also have another important consequence that de Quatrefages has stressed.[18] A single egg does not produce just one individual; it produces a large number of individuals, and in this way its ability to proliferate is greatly enhanced; the egg does not engender a single organism, but a whole generation.[19] This production of an entire generation, which the professor at the Museum calls *geneagenesis,* is particularly valuable for the lower forms of animal life that have little vitality or resistance and for parasites that have to run a thousand dangers before reaching a host in which they are able to live. In fact, it is primarily in these minor members of the animal kingdom that this prolific form of reproduction is most common. But while he demonstrated the practical importance of geneagenesis, de Quatrefages assigned to it its proper importance as an isolated phenomenon peculiar only to certain organisms. First of all, the purpose of geneagenesis is the same as that of metamorphosis. Both of these phenomena can reciprocally complicate and interpenetrate one another to the point that it is impossible to say where one ends and the other begins. Like geneagenesis, metamorphosis tends to augment the reproductive power of each individual. Such an increase can be obtained either by multiplying the number of organisms that a single egg can produce or by multiplying the number of eggs that each female can lay. But, because there is a limit to the size of the bodies of the female, the number of eggs it lays can be increased only if the size of the egg is reduced. All eggs contain two types of material, one from which the embryo is constructed and another that serves to nourish it. Obviously, the latter has less importance, and it is these nutritive materials that will experience reduction. On the other hand, no animal attains its full development without having undergone a great number of metamorphoses that are generally accomplished in the eggs of the higher orders of animal life. When the nutrient material stored in the egg is no longer sufficient to sustain the embryo through its full course of development, the embryo hatches before reaching its definitive form. It attempts to supplement the nourishment that it needs to assure the completion of its development by continuing, after hatching, the transformations that it began while still inside the egg. The larvae of insects are therefore only prematurely born embryos that have become capable of subsisting by themselves and can continue to develop when free-living. Among animals of the higher orders the growth and metamorphosis of the body advance hand-in-hand as essentially the same process rather than as a succession of stages as they do in insects and many other lower types of animals, but this does not diminish the importance

[18] A. de Quatrefages, *Métamorphoses de l'homme et des animaux,* (*Revue des Deux-Mondes*, 1855 and 1856 and vol. in-12, 1862).

[19] Perrier seems to be using 'generation' for 'population.' De Quatrefages is pointing out the fact that a large proportion of cloned individuals is being formed and that the resulting increased population of cloned individuals (*ramets* in botony) is adaptively advantageous in 'lower forms of life. (Trans. note)

of the metamorphosis. The basic process is the same in both insects and vertebrates, the only difference being the stage of life when the most obvious changes become apparent.

This added link between metamorphosis and geneagenesis is essentially morphological rather than teleological. In the course of his work, de Quatrefages compared the mode of growth of annelid worms with that of colonies of polyp hydra. The new segments of annelids are formed in exactly the same way as new polyps are added to an aqueous colony. In the annelids it is manifested in the formation of new segments that become an essential part of the animal's growing body, and these increments fill roles similar to some of those in the progressive growth of the bodies of higher animals such as mammals. Thus, the formation of the colonies of polyps can be related to a much better known, quite ordinary phenomenon, the growth of the body. The thing that is distinctive about these colonies is their arborescent form.

But in the case of the annelids, the addition of new segments often results in formation of autonomous individuals that are only products of the growth of the organism from which they become detached. The same thing takes place in colonies of polyps and leads to the formation of new colonies: this is what is known as *digenesis*. The growth of higher animals is always complicated by morphological changes, and the same is true of the annelids. The new individual that is formed can differ in notable ways from its parent, as it does in the case of the *Autolytus* and the *Syllis*. It is exactly the same in the relationship of aggregated to solitary salpas, the jellyfish to polyps, and in all cases in which there is *alternation of generation*.

As de Quatrefages said,[20] 'all agamic generation is related to true growth. This phenomenon is manifested as much by an *augmentation of the size of parts*, as by the *multiplication of these same parts*. In the latter case, it often happens that each added part reunites an assemblage that almost makes it a separate individual. In the annelids, for example, when the body attains its greatest extent, each ring possesses its own central nerves, an apparatus for motion, a vascular system, a large digestive poach, and reproductive organs, all of which existed in the preceding ring and will be in the one that follows. With only one more step each ring will be able to subsist on its own. All it really lacks is a mouth and sensatory organs. In the *Syllis*, the polychaetes, and the *naïs*, such a mouth has opened, and it is true that these organs appear on a special ring, but that ring is formed in exactly the same manner as the others. All the rings placed behind this accidental head obey it. A new individual that is formed has its origin in an assemblage of phenomena that differ in no way from those of the growth observed in any other members of the class. As Desor observed, there is no fundamental distinction between these phenomena and the gemmation of hydra and the strobile, or the segmentation of a single individual as described by Sars. The form of the species and the laws governing its growth are enough to explain the apparent differences. Thus one passes by

[20] *Metamorphose de l'homme et des animaux*, p. 268.

imperceptible variations from the simple growth of mammals to budding, and all this leads us back to the important conclusion that budding and asexual reproduction are basically only a *growth phenomenon.*'

For de Quatrefages, a mammal's body is an assemblage of individuals that have emerged from the egg of a *Syllis*, a polychaete, or naïs. The union of polyps forms a colony, and all the medusa that detach themselves from it are therefore equivalent.

'Once it is seen in this perspective,' he continued, 'we can understand very well why asexual generation cannot go on indefinitely. The growth has limits that are fixed in advance. If the budding is only a form of growth, its extent must necessarily be limited. It is not sufficient, therefore, to perpetuate a species. From then on, another mode of generation becomes necessary. No kind of animal life can escape this requirement. Thus the periodic return of sexual reproduction has a useful purpose, and Richard Owen's conclusion that the active elements of sperm have a limited fertilizing power finds added support that is independent of any hypothesis. Cuvier, Dugès, and, for other equally strong reasons, de Quatrefages, often included colonies of lower animal forms with what we take to be individual animals of a higher form, but just as Dugès had given Cuvier's idea new importance by showing its application to comparative anatomy, de Quatrefages gave unexpected support to Dugès' theory by successfully applying it to the more complicated phenomena of reproduction.

Henri Milne-Edwards proposed to set up, as Richard Owen did, a general theory for reproduction in which he sought to establish a strict parallelism between the phenomena of alternating generation and ordinary processes of sexual reproduction. He concluded that, far from being an exception, the phenomena seen in the development of the salpa and medusa are the general rule. All animals begin as a simple living vesicle that has been given the name *protoblast.* This protoblast is most often contained in the egg where it is the germinating vesicle. It generally terminates its short existence there, but it can also lead an independent life. Such is the case of the ciliated embryo of the *Distoma*. Before dying or disappearing, the protoblast produces by budding a more complex organism, the metazoan: it is the hydra polyp in the case of the medusa, the solitary salpa in the sea squirts [tuniciers], the blastoderm in the vertebrates. The metazoaires also tend to have only a brief existence: like the protoblast, they ordinarily disappear and, also like it, produce the definitive animal, the animal charged with perpetuating the species by way of sexual generation, the *typozoan*. The protoblasts can multiply under their own simple form and produce, as a result, one or more metazoans. The metazoans can produce several typozoans or they may produce only a single one with which they are sometimes confused. It is this more or less greater ability to reproduce, shown by the successive stages of the subjects of this series, that is responsible for the observed differences in the development of the animals. We should no longer be astonished by a phenomenon that is absolutely general.

In comparing the various theories that we have just laid out as an explanation for these phenomena, one will no doubt be surprised to see the wide differences in the views of their authors. For a physical scientist, the basic starting point of all theories is a simple phenomenon for which the governing conditions and laws have been rigorously established. One explores its diverse modifications through the widest range of possible conditions. All physical scientists are in agreement on this point, and we could add that they are also in agreement with chemists and astronomers on the goal of their theories. In a word, naturalists who cultivate the physical sciences in order to explain a complex phenomenon strive to show how it is related to another very simple phenomenon, the details of which are well known. To do this, one must strip it of all the accessory features that tend to modify it, so that it can be seen in its most elementary form. For example, astronomical phenomena can be related to the simple laws governing the gravitational attraction of bodies, and the science of astronomy is little more than an application of the basic law that states that bodies are attracted to each other by a force that is proportional to the product of their mass and the inverse square of their distance from one another. In the same way, all the phenomena of acoustics and optics are related to the laws governing the motion of a pendulum in that the theories of optics and acoustics are developed from the equations for vibration and waves. All the various transformations of thermal energy are derived from a simple principle: the heat generated when the motion of a body is suddenly arrested can be determined from an equation that establishes the equivalence between the mechanical and thermal energy. All electrodynamic phenomena are similarly governed by the attraction of one polarity of current to another, and the principles of electrodynamics are based on an equation as simple as the preceding ones. Thus we know enough to extend these principles to all the branches of the physical sciences, and scientists are in perfect accord on what it means to *explain*. All natural phenomena can be related to simple basic principles the laws of which can be determined experimentally, and one strives to learn how these phenomena are modified under all the diverse conditions one can imagine. That is the method of the experimental sciences, and it is to the credit of men like Bichat and Claude Bernard that they have shown that by applying this rigorous approach to physiology we can relate the elements of anatomy to the most fundamental physical properties.

Naturalists, on the other hand, appear to have very diverse ideas of what constitutes an explanation for phenomena of this kind. When they propose a theory, they seem to pursue very different goals. Steenstrup, in his theory of alternating generations, and de Quatrefages in his theory of geneagenesis were trying primarily to determine the purpose of the phenomena they described, and in that sense they followed Cuvier who would not accept any explanations that were not based on the principle of final causes. Leuckart, in laying out his theory of polymorphism, and Van Beneden, in developing his ideas about digenesis, simply stated that the phenomena that were at first thought to be exceptional are much more common than they had realized. They related these phenomena to others that are simpler and more general but are limited to one part of the animal kingdom and remain largely unexplained. Richard Owen confined himself to looking for a hypothesis that could relate two categories of phenomena that were considered distinct. This is in contrast to de Quatrefages and Milne-Edwards who demonstrated that a group of phenomena that are thought to be characteristic of certain organisms may be found in more or less modified form throughout the entire animal kingdom. They used phenomena that are observed in the higher vertebrates as a basis for comparison and related them to those that are present in lower organisms. They tried to use complex phenomena such as sexual reproduction and the embryonic development of higher animals as a basis for comparison of equivalent phenomena observed in lower animals. The course followed by these two illustrious French naturalists is exactly the reverse of that followed in the physical sciences. These differences may be an inevitable consequence of the fact that in attempting to understand forms of life that are more or less similar to our own, we have taken ourselves as the most perfect model of organized beings. We have tried to find organs, functions, and actions similar to ours in other vertebrates, and, confident that we know ourselves, we propose explanations based on analogies that we perceive between ourselves and the objects of their studies. In the hypothesis of the fixity of species, this was probably the most reasonable way to approach the problem.

When applying the concepts of inheritance [descendance] the problem is just the reverse: the method of searching for explanations is based on the methods of experimental science. We are no longer the model on which all is constructed and to which all must be related. Instead, we are the beings that must be explained, the end toward which the theory is directed and the most complex of the enigmas to be resolved. The explanations should be more than simple comparisons and generalizations; they must establish cause and effect relationships. This is especially important when dealing with the complex phenomena of alternating generation, digenesis, geneagenesis, and parthenogenesis which can be explained only by starting with the reproductive properties of the simplest beings. Once explanations have been found for the latter, the question becomes one of knowing in what measure these same explanations can explain the phenomena of development that are observed in the higher animals and man.

But one can utilize such methods only if the living being has first been reduced to its basic elements and the characteristics, properties, and faculties of the simplest forms of life have been determined. In order to do this it will no doubt be necessary to learn a great deal more about cell theory, which seems to promise a major advance toward resolving the problem.

Chapter XVIII

Cell Theory and the Constitution of the Individual

Pinel: the membranes. – Bichat: the tissues; their general properties. – Dujardin: the sarcode. – Schleiden: the cells of plants. – Schwann: extension of cellular theory to animals. – Prévost and Dumas: segmentation of the vitellus of the egg. – Studies of the origin of cells and anatomical elements of organisms; significance of the egg – Definition of the cell; protoplasm and the plastids. Constitution of the simplest individuals. – Animal colonies; numerous transitions between colonies and the individuals of higher orders. – Isadore Geoffroy Saint-Hilaire: life in colonies as a sign of inferiority. – De Lacaze-Duthiers: difference between the invertebrates and vertebrates. – General theory of the individuality of animals.

Since the beginning of the nineteenth century a growing number of philosophers, naturalists, and medical doctors have referred in their written work to living matter, organic molecules, animated material, organs, and tissues, but nowhere are these terms clearly defined. In the case of higher animals, one can distinguish the flesh, bones, fat, nerves, tendons, blood vessels, membranes; what do these substances consist of? Knowledge of this subject is limited to the comments about fibers that anatomists make in their descriptions of the muscles and nerves.

An eminent physician, Pinel, tried to apply the methods used by naturalists to a classification of maladies and found that this revealed a relationship between the progress of various kinds of inflammation and the nature of the membranes in which they are situated. This aroused the interest of doctors who wanted to learn more about these membranes and their relationships to the various parts of the human body. Bichat tried to address this need in his *Dissertation sur les membranes et leurs rapports généraux d'organisation* (1780), in his *Traité des membranes* (1800), and especially in his *Anatomie génerale* (1801), which appeared only a year before his death. In the first of these works, the young anatomist pointed out the similarities and differences between the membranes in various parts of the body and showed more clearly than earlier workers that similar membranes are found in very different parts of the organism. He classified them according to their structure, function, and external form. Three years later, he extended the method he had followed in this work to the full assemblage of organic systems. He devoted his studies of anatomy 'to identifying the attributes of each of

A. McBirney, S. Cook and G. Retallack, *The Philosophy of Zoology Before Darwin: A translated and annotated version of the French text by Edmond Perrier*,

the simple systems that combine in diverse ways to form our organs' and led the way for physiology, pathology, and therapeutics to gain a more exact knowledge of the properties of these 'simple systems' as they are in their natural state. General anatomy thus became a new science that was given the name *histology*.

'All animals,' he said,[1] 'are assemblages of diverse organs each of which has its own function and contributes in its own way to the survival of the whole. It is these particular mechanisms in the general apparatus that constitute the individual's entity, and they are heat, light, hydrogen, oxygen, carbon, nitrogen, phosphorous, etc. Similarly, anatomy has its simple tissues that, in sundry circumstances, form organs. These tissues are:

1. Cells
2. Nerves of animal life
3. Nerves of organic life
4. Arteries
5. Veins
6. Respiratory tissue
7. Digestive tissue and its glands
8. Bones
9. Medulla
10. Cartilage
11. Fibers
12. Fascia fiber-cartilaginous
13. Muscles of animal life
14. Muscles of organic life
15. Mucus membrane
16. Serous membrane
17. Synovial tissues
18. Glands
19. Dermal tissues
20. Epidermal tissue
21. Hair tissues

'Those are the organized elements making up our bodies. Whichever and wherever they may be, their nature is always the same. Just as in chemistry, the elements do not vary at all, whatever may be the compounds that they combine to form.'

Each of the various *tissues* that form our bodies has a particular kind of organization designed to make its own special contribution to the general life of the individual, but are there analogous assemblages of those elements in different living beings? Are these same tissues found in all animals? Are they, strictly speaking, the fundamental elements into which living bodies can be resolved? These are questions that the microscope will soon resolve.

---

[1] *Anatomie générale*, Introduction p. lxvi. Ed. Blandin, 1831.

For Bichat, life was a basic property of tissues, and the various ways in which it manifests itself are the consequences of the different ways these tissues are arranged. Even at the time he lived, naturalists already dreamed of relating tissues to something less complex. Oken thought that a small spherical mass of gelatinous material, the primitive mucus, or *Urschleim*, constituted the entire body of one of the simplest forms of life, the infusoria. He even visualized the higher organisms as aggregates of infusoria. At one point, the work of Ehrenberg led scientists to believe that the so-called simplicity of the infusoria was just an illusion and that the structure of the microscopic organisms was almost as complex as that of higher animals. Dujardin, a professor on the Faculty of Science at Rennes, was the first to definitely establish, in 1835, that life could be allied with a simplicity of organization of the kind proposed by Oken. He gave the name *sarcode* to an amorphous, living substance that by itself composed the body of a rather large number of lower forms of life. Despite the positive evidence that Dujardin gave for the existence of sarcode, a substance capable of living on its own, it attracted relatively little scientific notice.

Nevertheless, microscopic studies of plants have shown a remarkable unity in the structure of these organisms. It was long known that their tissues presented a multitude of more or less similar vacuoles that were often referred to as *cells*. In 1835, Johannes Müller identified a similar structure in a crystalline choroid, and in fatty material in the spinal column of vertebrate embryos. In 1838, Schleiden emphasized the importance of cells in the organization of plants when he showed that one could think of them as associations of cells. At the same time, he described the meaning of this term. According to him, the plant cell is a spheroidal cavity the wall of which is generally resistant and encrusted with cellulose containing a semi-fluid and disposed around a small central mass or *core* containing one or more *nuclei*. Similar entities [éléments] have been carefully described in animals. In 1839, Theodore Schwann, impressed by the simplicity of Schleiden's theory, compiled all that was then known about the existence of cells in animals and concluded that all animals are formed of cells that differ from those of the plants only in that they are ordinarily thinner and their enveloping membrane has greater plasticity. It was his view that these cells were formed spontaneously, either in the interior of other cells or in an amorphous substance interposed between existing cells.

Given the definition of the cells adopted by Schleiden and Schwann, one could not fail to be impressed by the structural identity that the eggs of most animals shared with these elements. The egg was essentially a cell. In 1824, Prévost and Dumas showed that the first stage of development consists of repeated segmentation of the contents of the egg. Bischoff and Reichert showed that the cells making up the bodies of animals are derived from these segmented spheres, and in 1844 Kölliker proposed the principle that, contrary to Schwann's view, 'cells are never developed spontaneously; on the contrary, all the elementary parts of the future embryo, as well as all the living elements of the adult animal are direct descendants of a unique primitive element, the egg.' Thus, animals are assemblages of cells that come one from another either by division or by budding, so that each of them can be traced through a series of generations back to the egg.

How can this proposal be reconciled, in its absolute form, with Dujardin's observations of animals formed solely of sarcode? At first, a great number of eminent anatomists thought that was impossible, but the difficulty was simply in the way that Schleiden and Schwann had pictured the primitive cell [élément anatomique]. Several studies ended up showing that of the three constituent parts of the cell, the *enveloping membranes*, the *nucleus*, and the *matrix* [contenu], only one was essential, the matrix.

The cell sometimes appeared to be reduced to its membrane and nucleus, but then all its vital functions have come to an end; it is dead. The matrix is thus the real living part of the cell. It was given the name *protoplasm* (Max-Schultze). But this protoplasm, by its constitution and its properties, is identical to the sarcode of Dujardin.

The sarcodic creatures could thereafter be thought of as being formed from one or more cells lacking an enveloping membrane, as is the case with many cells of the higher animals. They conform to the general rule that the cell can be defined only as *a mass of protoplasm or sarcode of limited size endowed with independent life, ordinarily with a nucleus in its interior and capable of isolating itself by means of a more or less resistant membrane*. The cell, so understood, is what Hæckel called a *plastid*, a simple term that we can now adopt, even though it is relatively recent.[2]

Living protoplasm is known only in the cellular state, that is to say, as limited masses with dimensions and forms that are also extremely variable and that can be considered as so many individuals. There are no confirmed examples of cells forming spontaneously either at the expense of free organic material or in an already organized matrix [milieu]. Most histologists seem to agree with Kölliker's proposition, and the classical studies of Pasteur have shown that in all cases where cells were reported or groups of cells were said to have formed spontaneously outside another organism the observers were victims of illusions. Thus all cells have been produced from other cells.

An isolated cell can produce other cells that separate from one another as soon as they are formed. This is the case with the simplest forms of life. But a single cell can also give birth to others that are destined to remain closely associated, and this is what happens in all forms of animal life ranging from sponges and polyps to man and in all plant life other than the monocellular cryptogams. All living beings are thus associations of cells, a fundamental proposition that is the basis of histology. Claude Bernard clearly demonstrated its importance in general physiology.

Even in their most complex associations, the cell that is part of a living being never loses its independence completely. Each one lives autonomously and the various physiological functions of the animal are nothing more than the result of

---

[2] For clarity's sake, we refer to 'plastid' and élément anatomique' as 'cell.' See Glossary. (Trans. note)

acts accomplished by a certain group of cells. It follows, moreover, that the entire physiology of animal or plant life is simply that of its constituent cells. If one could count the cells of an organism and if one knew their respective positions, properties, and relationships to one another, one would not only know all the functions of the organism, but one could also trace its embryonic development and predict the fate that awaits it. Thus, according to our present scientific knowledge, the cells are the anatomical elements the initial properties of which dominate all organic development, and studies of them promise to form the basis for all general theories of life.

All organisms now start life as a single cell – as the *egg of an animal or plant* – and have an embryonic development that progresses from the simple to the composite. The organism, which must develop from a single egg, displays increasing complexity. Scientific experiments indicate that in order to reach an understanding of the development and reproduction of the higher animals one must first determine all the traits of the lower organisms and gradually proceed to more highly perfected vertebrates. That would seem to be a common-sense rule, but since the vertebrates have long been the only animals whose organization was the subject of serious research, their embryology was the first to be studied, and naturalists have always strived to relate all the embryonic phenomena to it, much as there have been attempts to relate the phenomena of alternating generation to it. As a result, all the concepts bearing on embryology in general are burdened with a misleading kind of explanation.[3]

A general rule immediately emerges if one takes a logical approach and tries to determine the course of development of the most primitive types of sponges, cœlenterates, echinoderms, worms [vers], and arthropods [articulés]. The egg hardly ever leads directly to an organism similar to that from which it is derived; it first produces a very simple being. In the sponges and Cnidarians [hydroméduses], it is the first individual from which a colony is established; in the corals and echinoderms it is an organism that lacks the tentacles, arms, or appendages and will become the central part of the adult animal. In worms, it is what is called a *trochophore*. In the arthropods it is the *nauplius*. The trochophore and the nauplius simply represent the first segment [anneau] of the animal's body in the course of its formation. *This first segment always forms a part of the head of the adult animal* and sometimes the entire adult itself. From the point of view of its mode of formation, it corresponds exactly to the first individual of a colony of polyps, to the central part of a radiate animal [animal rayonné], the only difference being that it remains mobile, while the first individual of a colony of polyps quickly attaches itself to the substrate. The first polyp, the trochophore, and the nauplius also correspond closely in the roles that they will play in the course of the development of the animal:

---

[3] We possess numerous works on human embryology, but until now only one work on comparative embryology has been published, that of Balfour, which appeared in 1881. That work presents more than one proof of what we propose. In my *Colonies animaux* I have tried to develop the method that is described here.

by more or less irregular budding, the first polyp and its descendants will form the arborescent colony of which they become a part. By a peripheral budding, the central part of the radiate animal succeeds in producing an adult animal. By regular budding in a single direction, the trochophore and nauplius will make up the chain of segments that will form the body of an annelid worm [ver annelé] or of an arthropod. The only difference between the animals formed from segments placed end to end and the branching colonies of polyps is the direction in which the budding takes place.

This is what Charles Bonnet meant when he compared the organization of the tapeworm to that of trees. He pointed out that each segment of that animal could be considered a distinct individual, and he saw[4] a close analogy between the reproduction of lost parts of earth worms and the budding of plants.[5] Cuvier, however, had interpreted the comparison in the opposite way when he interpreted the sea pens and colonies of polyps as animals with several mouths. This is also what Dugès had concluded, and it prevented his *Theory of organic conformity*, which was so fruitful when applied in comparative anatomy, from lending itself to a complete systemization of the embryonic phenomena. We have seen how de Quatrefages attempted to achieve this systemization, but there again the illustrious scientist, having taken man as the basis for his reasoning, was only looking for analogies and failed to offer an explanation in the sense in which we now understand the word. If one follows the methods of the physical sciences, as Bichat wanted to do, this explanation must be deduced from the very properties of cells living in an isolated state. These properties are the following: first, the cells, in conditions suitable for *nutrition*, grow for a certain period of time. Second, each kind can reach only a certain maximum size beyond which they divide to produce new cells similar to themselves; it is this that we call *reproduction*. Third, the cells are subject to the influence of the conditions in which they are placed; their external form and physiological properties can be modified by circumstances; and in this way the cells show a certain *variability*.

The associated cells born from eggs preserve the same essential properties of nutrition, reproduction, and variability that are observed in the isolated cells. Moreover, they remain largely independent of one another. But owing to the large numbers of them that are closely associated, each finds itself in somewhat different conditions, lives in a manner all its own, and has special external characteristics. A division of physiological functions among the diverse cells soon develops and contributes to assuring the existence of the entire community [organism]. This division of functions makes the cells so specialized that the breaking up of their association necessarily brings about their deaths. This is how the *individuals* (and their *organs*) that result directly from development from the egg are made up.

The buds put out by these individuals produce complex aggregates, the members of which Dugès referred to as *zoonites*. They interact with one another much in the

[4] Bonnet, Considerations sur les corps organizes *Œuvres* vol. III: 226.

[5] Bonnet, *ibid.* proposition 255.

same way as the cells of which each is composed. Under certain conditions these second-order individuals take on new forms, accomplish special functions, and can either separate from each other or remain permanently united. The different phenomena known as *alternation of generations, digenesis, geneagenesis*, etc. are nothing more than the result of the early or late separation of the second-order individuals that differ to various degrees from one another.

When zoonites have not separated, the assemblage of united individuals constitutes an organism that we refer to as a *colony*, provided, of course, its members are clearly distinguishable from one another and appear to have preserved a large degree of autonomy. The term *individuals* is then used when the constituent zoonites are less clearly separated or when they all seem to be dominated by a single will that does not appear to reside in any particular one of them. One can see from this how vague the term individual is when one can pass at will from the cell to an aggregate of cells and from this simple aggregate of cells to an association of aggregates that resembles it. The boundary between what is called a *colony* and what is called an *individual* is at best somewhat arbitrary.

Isadore-Geoffroy Saint-Hilaire was impressed by these transitions from colonies to individuals that have attracted so much attention in recent years. In his admirable *Histoire naturelle générale des règnes organique*,[6] he used the word *community* instead of the more common term *colony*, and he emphasized in this way the similarities between these communities and what are ordinarily referred to as individuals.

'Like individuals,' he said,[7] 'the community has its abstract unity and collective identity. It is a union of individuals, often in great numbers, and yet it can and must be considered a single unit, a single but composite being. And its identity is more than an abstract concept; it is a tangible entity consisting of organized and reciprocally dependent parts, all of which contribute to the same assemblage, even though each one may be more or less clearly delineated. All are members of a single body, although each constitutes an organized body that is a miniature of the whole. Although the community as a whole enjoys a real, distinct identity, it is also true that the individuality may account for the fact that each being is very distinct from the others.

'In this way, every community combines two entities, two lives, two individualities so to speak, superimposed one upon the other… and in the final analysis, the definition we have given for the community can be summed up as an individual composed of individuals, or better, as individuals within an individual.

'Like a family or society, the community can be very diversely constituted. The anatomical fusion and hence the physiological solidarity of the united individuals can be either limited to certain points and vital functions, or it can extend to almost the entire combination of organs and functions. All intermediate degrees can also be represented, and one passes by imperceptible degrees from organized beings in

---

[6] *Histoire naturelle générale des règnes organique*, vol. II, p. 284.

[7] *Ibid.*, p. 295.

which the associated lives are almost entirely independent with distinct identities to others in which the individuals are more interdependent and intertwined, and to still others in which the parts are compounded into a common life in which all the individuals are more or less completely integrated into the collective identity.'

Judging from this admirable comparison of the community and the individual, one would think that Isidore Geoffroy Saint-Hilaire could offer some insight into their relationships. For example, he could have shown how the hydra polyps weld themselves together to produce medusas or how the segments of annelid worms and arthropods fuse and modify themselves in order to fulfill functions that are useless to one of them independently but indispensable to the existence of the assemblage as a whole. He could have explained how the varied phenomena observed in the communities offer an explanation for the formation of the complex organisms toward which it seems, according to his own words, they lead us step by step. One would like to hear him say that the history of the communities [colonies] is a series of spontaneous experiments prepared by nature so as to enable us to perceive the means by which it produces the higher organisms. But he says nothing about this. No conclusions are drawn from the experiments he describes. It is by the coalescence, welding, and more or less complete fusion of their constituents that the colonies become higher organisms. Instead of placing the colony high in the scale of nature, as had Dugès earlier, Isidore Geoffroy Saint-Hilaire interpreted this coalescence of individuals as lowering the status of the colony.

'In a group of *composite* molluscs or a colony of polyps,' he said, 'one easily sees the individuality of each of the *constituent* molluscs or polyps and how their individualities prevail over the character of the collective whole: in the tree, they balance one another or the individuality of the whole may even begin to prevail; while in the sponges, the influence of their individuality is really only theoretical... it was quite difficult to identify individuals in a vegetative colony; the numbers making up the mass of the sponge colony are not only beyond calculation; they cannot even be estimated. It is literally indefinite.'

And somewhat later:

'Communities [colonies or clones] are normally observed only in plants, a realm in which unitary life is not the rule and in the lower groups [taxa] of animal life. In order to find examples in the higher ranks of animal life, including man, one must resort to teratology, and there even the community is almost always reduced to the union of two individuals that, in most cases, cannot prolong their existence beyond the fetal stage of life.'

Thus the connecting strand is completely lost, and the question is no longer relevant. One clearly sees the unity of the community established piece by piece in the lower branches of the animal kingdom by the fusion of originally distinct individuals. But the fact that a relatively elevated organism can come about from the unification of a certain number of simpler organisms is completely neglected. Likewise, it is hard to imagine that this organism, that we call a vertebrate or arthropod – so completely of a whole and essentially indivisible – could have been produced by a process analogous to that which creates a siphonophore or a medusa from a colony of *Hydra*.

The contrast between the lower organisms capable of living in colonies and the higher animals that tend to live as separate individuals was already clearly recognized in the doctrine of Isidore Geoffroy, but this view was expressed especially well in the fine lectures presented in 1865 at the Museum of Natural History by one of the scientists who has carried out numerous studies of coral colonies, Professor Lacaze-Duthiers, founder of the laboratories of Roscoff and Banyuls.[8]

In one of these lectures, he first traced the main traits of the organization of the invertebrate animals then expressed his views about them as follows:

'A second impression one gains from studies of the invertebrates is that the individuals are exceedingly complex. In almost all these animals, what one ordinarily calls an individual is nothing more than an assemblage, a colony of small, more or less distinct individuals referred to by the general name *zoonites*. In order to form the complex entity, these zoonites join with one another, either in a linear series or in a mass, depending on whether it has two or three dimensions.'

The integration of the annelid worms, the arthropods, and the colonies of polyps is well described in the passage just cited, as it is in the *Mémoire sur la conformité organique*. The polyps of the colony and the segments of worms or insects are equally ranked as what can be called zoonites. The process by means of which the colonies develop to the level of organisms is the same as that which Dugès, Milne-Edwards, and Richard Owen have also pointed out. De Lacaze-Duthiers was also in agreement with Isidore Geoffroy, and his ingenious commentaries complemented the latter's thinking: In the lower types of animal life, all the individuals of a linear or irregular colony resemble one another and have considerable interdependence, but, little by little, an increasingly strict solidarity is established as a consequence of the division of physiological labor. 'In a colony of fresh-water hydra, for example, the individuals are bound to one another only at their lower extremities, but all the extremities with tentacles are free to function separately. The diverse species of *Clavelina* that belong to the class of sea squirts now considered chordates live together on radial prolongations that can be compared to the branches of strawberry plants, but their actions are otherwise free. The same thing is seen in other kinds of composite sea squirts; each of the colonies is enclosed in a single fleshy envelope with only one opening through which the creature can discharge waste, and there is already less independence in the vital functions. The siphonophores have colonies with a curious composition. Their zoonites are specialized in a particular way in which certain of them, having the form of elongated filaments that end with a vent-hole or a kind of harpoon, are designed for fishing. They seize bits

[8] The text of these lectures, published in the *Revue des cours scientifiques* does not bear the name of the professor, but I had the honor of being at the Ecole Normale Superieure at that time and was one of a group of students who were very attentive to the eminent author of the *Histoire naturelle du corail*. If my memory can be trusted, the edited version in the *Revue du cours* gives, at least in its basic form, a good account of Lacaze-Duthiers' thinking.

of food and pass them on to other zoonites, each of which has the form of a simple vesicular cavity or gastric horn. Other zoonites provide locomotion, and certain special ones have the function of giving birth to new individuals.

Further on, de Lacaze-Duthiers stresses the particular facility that the linear colonies contribute to the solidarity: 'In most linear colonies there are forced relations between a zoonite and its two neighbors, and these close relations can modify its form more or less completely. In massive colonies, this relationship is less absolute. We should also expect to find zoonites that differ very little from one another. This tends to verify the observation.' This last statement may have been a little exaggerated, and one could question whether the close contact that a zoonite in a linear colony has with its neighbors could have a strong influence on its form, but it is a matter here of phrases extracted from a lecture where the precision of language tends to be subordinated to the need to make an impression on the minds of the listeners. The more extensive perfection expected for some kinds of linear colonies is no less developed, and one of the important results of development is even indicated: 'Although each zoonite ordinarily possesses a nerve center, it must be noted that in the higher invertebrates, there seems to be a tendency to concentrate this nervous system in the anterior part of the animal.'

A tendency for organs that were initially disseminated to become concentrated in individual zoonites had the effect of stabilizing them and became one of the characteristics by which the higher organisms are distinguished from simple colonies. It may seem natural today to see the high individuality of the vertebrates as the final result of this concentration, but the work of Geoffroy Saint-Hilaire and Dugès helped prepare us for some of these ideas, and the anatomical, physiological, and embryonic research that followed left no doubt in the minds of even the most irreconcilable members of all these schools. In 1865, however, the proof that the vertebrates could also be broken down into such zoonites was still far from being established, and de Lacaze-Duthiers, instead of seeing the vertebrates as the culmination of the long series of invertebrate animals, was completely opposed to Lamarck's representation of the two sub-branches he had established in the animal kingdom.

He said that 'it is only the nervous system of the vertebrae that clearly differentiates the vertebrates from the invertebrates. *In many relationships, the latter are completely different from the former.* This *almost complete separation* has been cited by the obstinate critics belonging to the so-called *philosophical* school, including Geoffroy Saint-Hilaire in France and Goethe and Oken in Germany, and needs to be developed by further studies.

'One of the first things that must be recognized is the different distribution in the vertebrates and invertebrates of this mysterious thing that some say is the cause and others the effect of what we call life... If one thinks of life as a cause, an action having its origin in such and such center of organization, and if one is allowed to represent life as a quantity that will be more or less large according to the strength of the effect it produces, we can say that in invertebrates life seems to be distributed equally throughout all the parts of the organism. In the case of vertebrates, however, life is centered at a particular point in each individual or at least in a limited part of the body.

'If one wants to see life as an effect or result, one can follow the principle that I wish to propose by saying that among the invertebrates, this does not appear to be the consequence of something associated with a particular part of the organism as it is in the vertebrates, or, to use an expression that is probably too strict for things of this kind, it seems to manifest itself in one or more special, distinct organs.

'An example may help make this clearer. If one cuts off a dog's foot, the animal suffers, but it continues to live. One can go just so far in this sort of mutilation without bringing an end to life, but one always arrives at a place in the organism beyond which one cannot go without extinguishing life abruptly. This special point where life seems to be concentrated, this *vital core*, to use Flourens' expression, is found in all vertebrates. One can also express the same idea by recalling the picturesque image employed by Bichat when he portrayed life as being supported by a tripod, the three legs of which are the heart, lungs, and brain. If one of these three is destroyed, the tripod falls and life ends.

'On the other hand, if one takes an insect or any other articulate, cuts the body into parts and even severs the head, life does not cease. Try any kind of mutilations and it is quite evident that death will eventually come about, but we do not find in this creature a point analogous to the vital core or one of the three fundamental organs found in the vertebrates, a point or organ which, if severed, will bring an abrupt end to life.'

Thus the vertebrates are represented here as the exact opposite of the invertebrates; there are fundamental differences between the two. Life is conducted very differently in the privileged sub-realm with which we associate our anatomical structure, and some zoologists still confuse, as Lamarck did, this sub-realm with that of all the other types of organisms. It is because of the exceptional centralization that one observes in the higher vertebrates that they are considered to be a special form of life, but the importance of this centralization was greatly exaggerated by Bichat, as was shown by the example of Flourens' chicken that lived for a month after its brain was removed. It shows the increasingly greater dominance of functions of the spinal marrow over those of the brain as one considers successively lower types of vertebrates. On the other hand, this centralization is precisely the same phenomenon by which communities gradually take on the character of individual organisms. As we have seen, this is already developed to a high degree in the arthropods, and the only difference in the case of the vertebrates is that they have advanced to a higher degree. Can this difference of degree allow one to overlook the relations that Geoffroy Saint-Hilaire, Ampère, Dugès, Goethe, Oken, Richard Owen, Leydig, and de Quatrefages pointed out between the segmented organization or the mode of development of the vertebrates and the segmented organization or mode of development of the annelid worms and the arthropods? Obviously not. If it were so, if the vertebrates were really formed from zoonites as the invertebrates are, if they differ only in a greater degree of coalescence of their zoonites, there is no longer any way to set them apart. The same law of evolution applies to the entire animal kingdom. Among vertebrates, just as among invertebrates, organic complexity has been attained by a more or less complete fusion of zoonites, that have budded on one another, and the first, to which the name *protoméride*[9] can be given, was the only direct product of an egg.

In summary, all these considerations must lead to a simple concept of the evolution of the individuality of animals. It is first reduced to a unique cell [plastid], the egg. The egg produces, by repeated division of its substance, more or fewer new cells. These new cells can separate as soon as they are formed and go on to multiply in turn in the same or different forms. This is what happens in a great number of protozoa.

The division of the egg may or may not be preceded by its intimate union with an element similar to itself [isogamy?] or in the form of a mobile filament [spermatozoan]. In the first case there is *conjungation*, in the second *fertilization.* Fertilization almost always precedes the division of the egg when the latter must lead to the production of the cells destined to remain associated. Its absence constitutes the phenomenon of *parthenogenesis*.

The cells that remain associated are not constrained to maintain a unique form. As soon as they differentiate, they form a relatively simple, poorly defined organism to which we shall give the name mérides.[10]

The mérides multiply just as cells do. Some reproduce directly from eggs; others give birth to new mérides, that can separate from their parent as soon as they are born and live independently. This is true of some of the lower sponges, the freshwater hydra, and certain lower types of worms. A part of the phenomenon of *alternation of generations* and *geneagenesis* is related to this mode of development – mérides playing a role that Van Beneden called *digenesis*.

Mérides born one from another can also remain united. They then form what have been called *communities* or *colonies*. The mérides of a single colony can assume various forms to accomplish different functions. Groups that are appropriate to these functions can detach themselves from the colony in which they were born and give rise to very remarkable examples of geneagenesis and alternating generation. This is what one observes in the alternation of generations of jellyfish and annelids. But all the mérides born from one another can remain joined together, modify themselves in different ways, and become so conjoined [solidaires] that they are inseparable. Their assemblage then constitutes a new organism having all the characteristics of an individual entity. This is true of all the higher animals that can be called *zooids* or *dèmes*, in that they can break down directly into mérides. It is first necessary to distinguish in them groups of mérides and zooids having the same attributes or particular functions just as the bodies of animals have several distinct regions. When the protoméride becomes sessile, it puts out buds of irregular

[9] From πρϖτον, first, and μεροs, part.

[10] See my *Colonies animals* pages 403 and 705.

arborescent or incrusting colonies in which there are a certain number of equivalent individuals are assembled around a common center to produce radiating organisms.

When the protoméride remains free and rampant, it has a bilateral symmetry, produces buds only at its rear extremity, and gives rise to segmented organisms of which the annelid worms, arthropods, and vertebrates are the principal forms. Thus, the different modes of symmetry that characterize the major organic types have a rational explanation, and it is no longer necessary to call upon direct intervention of a distinct creative will to account for it.

The production of the protoméride, the formation of mérides and zooids, all the phenomena of reproduction that require no fertilization, and all the phenomena of *metagenesis* can take place successively and form several more or less distinct stages of development. They can also take place more or less quickly, and often quite quickly, in order to complete their development before hatching [éclosion]. It is due to the differing degrees of this *acceleration of metagenetic phenomena* that animals seem to go through several types of development.

This acceleration reaches its maximum in the higher organisms of each group: certain jellyfish, some of the modern echinoderms, higher crustaceans, arachnids, insects, molluscs, and vertebrates all hatch from the egg with all the mérides that must constitute them and only undergo modifications that are just a matter of details, while most coelenterates, crinoids, most worms, and the lower crustaceans still possess at birth only a small number of mérides and often only one.

Thus a single theory unifies all the major features of the gradual formation and definitive structure of organized individuals. It is a simple matter to explain these individuals if one first tries to learn how they are developed and thinks of them as a *result*; it is much more difficult to define them if one considers them independently of all the forms that they go through and insists on seeing *primordial facts* in them. We find here the same opposition pointed out earlier between the clarity brought to the natural sciences by the hypothesis of transformism [evolution] and the hopeless confusion that is always caused by the hypothesis of fixity of living forms. It is a mistake to want to include in a single definition the *individual* as the higher groups of the animal kingdom show it and the floating forms that are so common in the lower groups; the individual no longer exists in the latter.

It is almost useless to point out that the theory of the formation of individuality, that we have just laid out, can be interpreted just as well as indicating the course that must have been followed by living beings in arriving at their present degree of development if life on Earth began with simple forms comparable to the cells. We can attempt to learn what these conditions could have been, but when we do we are reduced to conjectures. What conditions prevailed when the first cells were formed? Why does their formation appear to have ceased? Why have we remained unable until now to form living protoplasm from all the pieces? Those are the questions to which we have not even begun to see scientific answers. In a similar way, most fields of science have been unable to get at these questions of origin: astronomy does not know where the material came from and how it made up the stars whose development it studies. Physics does not know the cause of the various kinds of movement and their rhythms even though it has been able to use mathematical laws to relate the innumerable phenomena that produce weight, heat, light, electricity, magnetism, and simple forms of motion. Chemistry is still trying to understand simple compounds and the conditions under which the elements that seem so immutable could be formed. Biology, putting aside the question of the first appearance of life and living substance, remains in the same condition as all the observational sciences. It is enough for it to have acquired the knowledge of the elements [cells] that combine in various ways to form the living beings it studies.

Before the appearance of Darwin's book, all the information needed to construct this theory of the individuality of animals was already known to science. There is not one of his chapters that some naturalist could not at some stage have formulated in his mind. But all the facts were widely scattered, and it is only in recent years that they have been brought together.

Knowing the detailed make up of the individual and its probable mode of paleontological evolution, it is necessary to determine how this condition came to be manifested in each individual. That is the role of embryogeny, which we must render more precise for the enrichment of zoological philosophy.

Chapter XIX

Embryology

Epigenesis and embryogenesis. – Harvey: influence of the cellular theory. – The egg considered as a cell. – Theory of blastodermic leaflets, - Exaggerated generalization of results obtained from studies of vertebrates. – Embryology from the point of view of histogenesis and organogenesis. – Serres and transcendent anatomy. – Embryology considered as transitional comparative anatomy. – Arguments supporting this theory. –Embryological Classification; their shortcomings. – The embryology of an organism is its condensed genealogy – Embryological acceleration; perturbing phenomena that result from it. – Links between embryogenesis, general morphology, and paleontology.

Studies of embryology date back no farther than the day when naturalists firmly overturned the hypothesis that living beings were entirely contained in their primitive germ, that all their transformations consist of growth of its parts, and that organs that are initially invisible gradually become more apparent over time. For many years, this sterile hypothesis, to which the great names of Swammerdam, Malebranche, Leibnitz, Haller, Bonnet, and even Cuvier were attached, resisted all efforts to overturn it. As late as the first half of this century, its partisans were still struggling against the growing evidence. As early as 1652, Harvey had posed the principle of embryology in its basic form when he affirmed that all living beings arise from an egg. In truth, his insight was a simple but ingenious stroke of intuition. The aphorism: '*Omne vivum ex ovo*' [all life comes from the egg], could be shown to be valid only if one can establish the nature of the egg and find such eggs in all living creatures. Régner de Graaf, who died in1673, was the first to recognize the eggs of mammals in the fallopian tubes of the womb and to find the part of the ovary where the egg is formed, but he did not identify the actual presence there of the egg itself. It was not until a hundred and fifty years later, that von Baer established that it is precisely in the *Graafian follicle* that the eggs of mammals are created, but the parts of this egg were not satisfactorily correlated with those of the eggs of birds until Coste did so in 1834.

The discovery of spermatozoids by Hamm and Leuwenhœk served only to feed the discussions between the *ovulists* who wanted the germ to reside in the egg and the *animalculists* who wanted it in the sperm. Two contemporaries, Prévost and Dumas, have definitely established that the penetration of sperm into the egg is necessary for the development of the latter and constitutes what is essentially fertilization. In any case, as de Quatrefages observed with regard to the eggs of

A. McBirney, S. Cook and G. Retallack, *The Philosophy of Zoology Before Darwin: A translated and annotated version of the French text by Edmond Perrier*,

tube worms, as well as the constant development of the infertile eggs of bees and other *Hymenoptera*, and many other lines of evidence, fertilization is not indispensable to the beginning of the genetic process. Swammerdam had already noted that the components of the fertilized egg divide themselves into distinct parts, and Prévost and Dumas showed that this *segmentation of the vitellus* of the egg was the primary factor in embryonic development. It was soon recognized that this phenomenon is almost ubiquitous, but its importance became apparent only after the cellular theory was well established. Anatomists soon realized that the egg was nothing more than a cell, the first of the histological elements [cells] of the embryo and the progenitor of all the others. Kölliker also concluded that segmentation is only one form of cellular division, and along with Bischoff, Reichert, and Virchow, he maintained that the cells of the embryo, including all those of the adult animal descend by an uninterrupted series of successive divisions and are true descendents of the ovular cell. To Harvey's aphorism *Omne vivum ex ovo* was added that of Virchow: *Omnis cellula è cellula* [all cells come from other cells]. Actually, the second of these propositions includes the first. The simplest living beings can be thought of as consisting of a histological element, a unique cell, and reciprocally the cell or associated cells making up organisms that are themselves true living beings with an independent identity. Harvey's aphorism, as well as that of Virchow, amount to saying that it would not be possible to have spontaneous generation either within the living organism or outside it. In truth, one must consider this proposal very carefully and not exclude the possibility of transformation into well-defined cells of amorphous protoplasm, such as those that are sometimes noted in tissues undergoing formation. They go by the name *blastème* [syncitium, coenocyte]. This opinion has been supported by eminent histologists such as Charles Robin.

From now on, the principal task of embryology will be to learn how all the elements that concur to form the body of a human (or any other animal) develop from the egg and determine all the stages that the embryo goes through before reaching the state of a well- defined organism. This problem is linked to another: determining the purpose of the succession of different forms that the embryo must go through in order to reach the final form that is the goal of its development.

Well before the significance of the egg and its first stages of development could be understood, the phenomena were already being considered from these two different points of view. While some embryologists were trying to determine the mode of formation of the tissues and organs, others were concerned mainly with the general relationships between the successive forms of the embryos and those of the adult animals. It is now possible to see a connection between the results obtained through these two different approaches, but the two schools have left marks of their separate lines of research, and their influence can be recognized in the research of our contemporaries.

Humans, together with certain uncommon mammals and chickens, naturally served as the basic subjects that embryologists drew on for their studies of tissues and organs. In the beginning, embryology, like other branches of zoology, was following an essentially anti-scientific approach. Instead of proceeding, as in the experimental sciences, from the simple to the complex, the most complex phenomena were studied in an attempt to understand the most simple ones.

Gaspard-Fréderic Wolf (1733–1794) had observed that the intestinal tube of the chicken first appears in the form of a flat lamina [feullet] that doubles back on itself little by little until the edges finally become welded together. He assumed that the other systems of organs had a similar origin, and in 1817, Pander concluded that all the organs stem from three thin superimposed layers or laminae [feuillets]. Today, these three *germinative laminae* are the subject of much embryonic research. Pander referred to them as the *mucous layer*, the *watery layer*, and the *vascular layer*. Von Baer, under the obvious influence of theoretical ideas that had much in common with those of Bichat, added a fourth embryonic layer and divided them into two groups: an *animal group* consisting of the *cutaneous layer* and the *muscular layer*, and a *vegetative group*, comprising the *vascular and mucous layers*. Following the studies of Reichert and Remak, it is now generally agreed that there are three *blastodermal layers*: 1. the *ectoderm* or external layer that produces the epiderm, the nervous system, and its dependencies; this could be called the *sensory layer*; 2. the *mesoderm* or middle layer that constitutes the muscles and blood vessels and that he called the *moto-germinative* layer; and finally, 3. the *endoderm* or internal layer that produces the epithelium of the digestive tube and its associated glands and was given the name *intestinal-glandular* layer.

Relating all the embryonic phenomena to the successive transformations of these three distinct layers no doubt greatly facilitated comparisons of these phenomena in the various organisms. As a result, observers continue to devote all their efforts to finding these layers in the embryos of all animals so they can determine their modes of formation and their diverse metamorphisms, thus extending to the entire animal realm the results that had been furnished by the study of vertebrates alone. This generalization could not have been achieved without considerable modification of the original terminology. It is no longer thought that the embryos of most lower animals are made up of three flat, superimposed *laminations*; instead, two invaginations [sacs] enclosed one within the other, share a common orifice, and new tissues are formed between them in diverse ways. The comprehensive term *mesoderm* has been adopted for them. These two invaginations are not always present. The larvae of sponges, as well as those of most cnidarians [cœlentérés] develop parts comparable to an ectoderm and endoderm only at a late stage, so that no general embryological theory could be based on the three blastodermal layers of the vertebrates. Moreover, the problem is not to find more or less exact analogs of these layers in the entire animal kingdom but to explain why the earliest forms of embryos of most vertebrates consist of three laminar layers. The theory of layers proved to have useful applications to organogenesis and histogenesis. It made it possible to coordinate a number of observed facts, but zoological philosophy obviously has nothing to gain from a doctrine that, from the outset, regards as resolved the problem for which it should, in fact, be seeking the solution.

The extended horizons opened by this work gave embryologists a fresh view of general morphology and led them to explore possible relationships between embryonic forms and the adult forms of related animals.

The obvious resemblance that the tadpoles of frogs and other amphibians have to fish had already inspired Kielmayer's idea that, before reaching the adult state, the higher animals pass through successive forms similar to the permanent ones seen in lower animals of the same group. We have found this same idea in the writings of Autenrieth, as well as those of the German natural philosophers and especially in the works of Geoffroy Saint-Hilaire who used them most successfully to identify analogous parts in various classes of vertebrates. But it was Serres, one of Geoffroy's students and like him a professor at the Museum of Natural History, who did the most to demonstrate close ties between embryology, comparative anatomy, and even the external morphology of animals.[1] Following the example of the German natural philosophers whose thinking he sometimes shared, Serres accepted as an obvious principle that 'man's constitution is really a miniature world'[2] that reflects the entire history of the animal kingdom. This hypothesis, which could well be the final conclusion of his philosophy, is actually its fundamental basis. It is an *à priori* assumption upon which Serres tried to organize the factual evidence, and from the beginning, the doctrine he constructed on this foundation took on a certain aura of grandeur. Man being the culmination of the animal kingdom, his organs and physical development go through the same successive stages as the those seen in the genera, families, and classes that make up the scale of the animal kingdom. The history of the formation of human organs is a repetition in miniature of the history of animal organs in general. 'The succession of animals is nothing but a long chain of embryos, laid out step by step and ultimately arriving at man.'[3] Endowed with a limited duration of life, the lower organisms stop sooner or later along the path to maturity that is rapidly traversed by the human embryo. 'Stopped at one place, it continues on at another. There you have the whole secret of development, the fundamental difference that the human mind can grasp between human organogenesis on the one hand and comparative anatomy on the other.' And one might say that '*human organogenesis is a brief summary of comparative anatomy, just as comparative anatomy is the fixed and permanent record of human organogenesis.*'

In his academic discussions with Cuvier, Geoffroy Saint-Hilaire implicitly related the unified plan of structure in the animal kingdom to a single plan of development. It is this latter unity that, according to Serres, is the essential law of nature, 'so that the entire animal kingdom appears to be only a single animal that in the course of formation in diverse organisms is arrested in its development, sometimes sooner and sometimes later, and this determines (at the time of each

[1] See especially *Précis d'anatomie transcendante appliquée à la physiologie* 1842.

[2] Serres *loc. cit.* vol. 1, p. 95.

[3] Serres *loc. cit.* page 91.

interruption of its current stage of development) the distinctive organic characteristics of the classes, families, genera, or species.'[4] The history of the lower animals, monstrosities, and fossil animals is closely connected to the genesis of organs, and one can understand that, conscious of vast domains that he was dealing with, Serres embellished the grandiose science that he visualized with the name *transcendent anatomy*. However, the point of view that this ingenious professor of comparative anatomy adopted was not a very sophisticated one. His preoccupation with seeing man in everything prevented him from appreciating the variety in the animal kingdom and from recognizing the true relations that unite the living forms. It is strangely misleading to think that things in nature come about as simply as Serres supposed. If man's superior intelligence raised him to an incommensurable height above the animal kingdom and his brain can be considered a reflection, from the point of view of the nervous system, of the ultimate manifestation of organic evolution, the same is certainly not true of his other organs. The human digestive organs are less perfect than those of the ruminants, his respiratory and circulatory systems are less refined than the analogous organs in birds, and his nutritional organs have nothing that places them notably above those of many animals. His organs of sight, taste, and smell are less sensitive than those of many carnivorous mammals, and his hand, which has been the subject of so much poetic praise, has developed much less from primitive pentadactyl forms than has the foot of the antelope or horse. Thus, there is no evidence that human embryogenesis is the evolutionary culmination of the entire animal kingdom or that its anatomy could be the most highly perfected. At no stage of its development does the human embryo pass through the form of a true fish, reptile, or bird before reaching a more advanced level. That is the objection that embryologists have raised with respect to Serres' theory, and as a result his transcendental anatomy has been discredited.

There can be no doubt, however, that a large part of the factual evidence on which this theory rests is perfectly valid. For example, certain stages of the development of the circulation system of the fetuses of mammals recall those of reptiles. The initial structure of the cranium is not unlike that of fish. At that early stage, their face presents arches that are comparable to the branchial arches of fish. The first phases of development of their head and body are common to all vertebrates. On the other hand, the organization of very young amphibians resembles that of true fish. The embryos of birds are much more analogous to reptiles than to adult birds, and if one compares the positions of the principal organs with respect to the vitellus in the embryos of vertebrates and arthropods one is surprised to find that they are absolutely identical, whereas, they become quite different in the adults.

Every day, there are new additions to this list of well-established facts, and embryogenesis never ceases to bring zoologists new surprises. In addition to the marvelous phenomena such as alternation of generations, which we have seen to be of great importance, we find that the greatest number of Cuvier's medusae [acalèphes] begin as polyps. These two sorts of animals later become confused

[4] Serres, *loc. cit.* page 19.

with one another, and it seems that polyps could be thought of as medusae that have been arrested in their development. Johannes Müller studied the unique metamorphosis of echinoderms, and felt justified in comparing the transparent larva of the regular echinoderms to the [acalèphes]. Thompson thought at one point that he had discovered a small living sea lily on our coasts, but he soon realized that it was nothing but a larva of the comatulid [crinoides] which reproduces in this way during its youth. At that stage it is a lower form of a group almost all the representatives of which are now found only as fossils. Thus, present-day animals, in their youthful stages, can bring back to life forms that have long since disappeared, and in this way a link between paleontology and embryology that Serres predicted turns out to be quite feasible.

Although they may not have the same significance, the metamorphosis of flukes [Trématodes] and the tapeworms [Cestoïdes] appears to relate parasitic worms so well to infusoria that Louis Agassiz proposed dropping this class of microscopic beings which he considered nothing more than larva of higher animals. The manner in which annelids develop suggested to Milne-Edwards and Quatrefages the attractive ideas that we discussed earlier. Thompson, Nordmann, and other observers showed that all the lower crustaceans have a common larval form, the nauplius, that was first thought to be an autonomous organism, a special kind of crustacean. Many decapod crustaceans are born as true schizopods; for much of their life crabs preserve a normal abdomen before becoming short-tailed [brachyure]. Even more remarkable, Thompson discovered that the nauplius is also the larval form of barnacles, so that they are permanently removed from the mollusks and are assigned instead to the arthropods. Spence Bate showed that, after having been nauplius, barnacles take a form that closely resembles that of another crustacean, the cypris, and their further development is arrested at that point. Numerous closely related studies have established that all the gastropod molluscs on the one hand and all the bivalve mollusks on the other have a common larval form and that these two forms can easily change from one to the other. In their early stages, shell-less gastropods are indistinguishable from others, but their larva possess a shell and an operculum like those of ordinary gastropods. Quatrefages concluded from studies of the development of the Taredo [shipworm] that when this strange mollusk becomes an adult it first takes on the same larval form as the others bivalves and, like them, has a bivalve shell into which it can withdraw completely. Moreover, the magnificent studies of the tusk shells [Dentalium] by de Lacaze-Duthiers revealed the striking peculiarity of a mollusk that is intermediate between the gastropods and bivalves; its larva is initially rather similar to that of a worm but later becomes identical to that of ordinary bivalves. Lovén observed that the appearance of larva of the chitons is also similar to that of a worm. The mollusks that Serres likened to the fetus of a vertebrate that never loses its fetal membranes have the form of worms when they are very young.

Thus, embryology continues to make increasing numbers of contributions to systematic zoology. It often reveals the most unexpected relationships between groups that have no outwardly apparent genetic connection. Not only does one find oneself obliged to recognize the specific identity of beings that were mistakenly placed in different genera or even different families, but we have had to abolish what were thought to be entire classes of animal life. The most eminent naturalists affirm the impossibility of determining the place of a particular animal in the system without carefully following it from the earliest developmental phases of the egg to a stage where it becomes capable of reproducing itself sexually. This has brought us those beautiful monographs for which Quatrefages' *Histoire naturelle du Taret* has served as a model, and work of this kind by de Lacaze-Duthiers has continued to enrich French science for thirty years.

The word embryology has a very broad meaning. Asexual reproduction, alternating generation, and metamorphosis that takes place in the egg or after birth have now become a productive field of embryonic research. In discussing these phenomena, I have shown how close are the ties that unite them to the phenomena of development and how much light their study has shed on the manner in which organisms are constructed.

Embryology could not fulfill its potential importance until naturalists tried to systematize the results to which it has led. The explanation of the transformations that each organism undergoes in its individual development appears much too remote for one to be much troubled by it. We no longer shirk the necessary attempts of Serres but remain convinced that they are not completely abortive and noting that a better formula has been discovered, we apply the transient characteristics seen in embryology to classifications – despite Cuvier's criticism of them.

Von Baer was probably the first to publish a purely embryonic classification. The four modes of development that he distinguished within the animal kingdom enabled him not only to define Cuvier's main branches a little more closely, but the characteristic of the vertebrate division relative to that of the arthropods is so sharp that it is the only thing that can still be conserved today, and the subdivisions that he proposed for that division served as the basis for all the later improvements. It is there that, for the first time, the vertebrates that have an allantois were separated from those that do not, and he called upon the various dispositions of the umbilical cord, allantois, and placenta to distinguish the sub-classes and orders of mammals. We now know how well the various modifications of their placenta have facilitated the classification of mammals.

Von Baer's primordial groups were not adequately characterized. Van Beneden chose to define these groups by making use of the relations of the embryo and the vitellus. He called animals in which the embryo resides on the vitellus by its ventral side hypocotyledons or hypovitelliens. All the other animals that constitute Linnaeus' former major class of Vermes were called allocotyledons. It is evident that this last division, based on characteristics that are strictly negative is in no way equivalent to the other two. That alone is enough to show that at the time when Van Beneden conceived his system embryogenesis still had something to tell us.

Kölliker preferred to characterize his divisions by making use of the rather large part that the vitellus plays in forming the embryo. Finally, Carl Vogt proposed a system in which he took into account the characteristics used by Von

Baer, Van Beneden, and Kölliker, but he also introduced other characteristics borrowed from anatomy or drawn from the existence of a cephalic vitellus in the cephalopods.

It is certainly true that these proposed classifications have not been particularly successful, and the same is no less true of all those that have been based primarily on embryology. One would expect more from a science that had been able to make such beneficial modifications to the older methods and had introduced so many fruitful new biological concepts. How do we explain the misleading ideas that it seems to have entailed?

It will be noticed that in all the proposed embryonic classifications, including even the most modern ones, the importance of embryonic phenomena themselves has not been taken fully into account. From Bonnet until the time of Fritz Müller, naturalists have struggled in vain (with speculations all too general to be precise), to demonstrate that the development of the individual was nothing but the abbreviated development of its species. This proposition, which all evolutionists accept today and seems to have again earned embryology the title of transcendent anatomy, seemed very promising, but it has found no application in any of the proposed classifications.

In effect, it means that the embryogenesis of an animal results from at least four factors that intervene simultaneously to produce the series of observed phenomena. These factors are: (1) heredity, (2) embryonic acceleration, (3) the mode of nutrition of the embryo, and (4) the independence of the cells, tissues, organs, and organ systems.

With regard to heredity, an animal should pass, in the course of its development, through a series of all the forms that its direct ancestors went through in the course of time. As these ancestors have left descendents that were modified in various ways and others that reproduce the ancestral forms more or less exactly, it is evident that, if our proposition is correct, comparative embryogenesis should always permit us to discern the degree of relationship of the animals belonging to the same lineage. This alone should furnish the means to set up an authentic genealogical tree of the kingdom, to formulate laws of comparative anatomy, and to institute methods of classification that are truly natural. The characteristics it displays take precedence over all others.

All these conclusions are perfectly legitimate, but they are based on the assumption that nothing could intervene to disturb the series of morphologies that heredity leaves on the embryo and that, once formed, these same forms could not be modified. Neither of these assumptions is justified. All the forms resulting from changes that the ancestors of a given animal have undergone were necessarily capable of adapting themselves to an independent life, at least during the period when they continued to reproduce. Regardless of the stage when the envelope of the egg was broken, it would seem that the embryo would have been capable of continuing to live freely and be able to search for the nourishment essential to its ultimate development. We all know that it cannot be that way. If the successive forms of the present embryo represent ancestral forms, they have certainly been profoundly modified. Only the ancestral forms are important when one is comparing adult animals, and this is the main thing that classifications and anatomical analyses are striving for. Without them, these forms will offer only dubious evidence, and one cannot distinguish what is primitive in the embryonic form from what has been modified.

This distinction will obviously be facilitated if one knows the basic causes responsible for modifying the embryo. Among these are the three factors we have just recognized as having an important influence. First of all, it is evident that, if the embryo passes through all the phases that its species have gone through, it has greatly condensed their duration. To the extent that the generations of different forms succeed one another, the duration of these stages becomes shorter and shorter in a way that development requires almost the same period of time. Thus, the embryonic development must have been considerably accelerated. This acceleration has brought rapid modifications of animal forms much as it has for the changes that the larva of insects go through in successive stages of their growth. In the case of the larva of insects, these incessant transformations cannot be related to the activity of organs; the embryo remains in repose protected by the shell of the egg during most of its period of development. In any case, in a single zoological group, its hatching can take place at various evolutionary stages. For example, in the order of decapod crustaceans, the *Penœus* comes out of the egg in the nauplius state. Shrimp and most of the other decapods appear in the stage of zoea that, in the case of the Penaeus, follows that of the nauplius. These zoeas then take on the appearance of a mysis, and it is only in this latter form that scyllares [sand lobster], spiny lobster, and true lobster hatch. Finally, the mysis stage is passed in turn in the egg in the case of the hermit crab and crayfish that are born with all the characteristics of true decapods.

One can conclude from this that embryonic acceleration is far from being the same for all species of a given group. Its effects can be highly varied with one stage being prolonged at the expense of another, so that the latter persists when the former has become unrecognizable or even entirely suppressed.

Finally, if accelerations affect all stages simultaneously, and development proceeds (in some manner) toward the goal of attaining an adult animal as quickly and as economically as possible, then the course of development might be completely transformed. Consequently, entire phases of development could be skipped, the reproductive cavity and its contents may develop in diverse ways, the embryonic envelopes may or may not appear as a result of moltings 'occurring in the egg or from other causes – all this without the ultimate (adult) forms' differing from one to another.

On the other hand, the transformations and metamorphoses that the embryo goes through within the envelope of the egg represent a process that cannot be accomplished unless the essential cells have sufficient nourishment. The accumulated nutritional reserves that the embryo has in the egg enable it to develop at an accelerated rate. The more delayed the hatching the more nutrition is needed, and the limited nutrition available in such a restricted space will lead to important modifications of the mode of development. The retarded development of the mouths of vertebrates or the disposition of their earliest embryonic stages in widely spaced, open layers must be examples of this kind of modification. If one examines the characteristics on which the diverse embryonic classifications have been based, it is evident that the only ones that have proved useful are precisely those that result from the intervention of two perturbing aspects of embryonic development: embryonic acceleration, and the accumulation of nutrient material in the egg. It is quite clear, however, that such characteristics could not have more than a very subordinate importance. They could be effective only in the much higher groups in which adaptation to a narrow range of conditions will entail the appearance in their embryos of true hereditary organs responsible for nourishing the organism. This is how the allantois can distinguish vertebrates that have definitively adapted to an aerial existence from those that have not yet succeeded in doing so. The different forms of the placenta reflect rather well the affinities among various orders of mammals. But in that case it is not of the modes of development but the well-defined organs developed by a long elaboration that are as useful for classification as the feet or teeth of the adult animal. Thus, all the purely embryonic classifications have failed because they have drawn on characteristic mechanisms of development found in the most diverse types and processes resulting from the perturbations of the normal embryo rather than their essential phenomena. Before the hypothesis of evolution [transformisme] had revealed the true significance of embryology, many naturalists, doubtless from a quite natural reaction to the exaggerations of the naturphilosophs, had completely lost sight of the parallelism between the individual development of the higher organisms and the series of beings that start as the simplest forms and gradually achieve their full development. Since the doctrine of evolution has led us to treat the embryogenesis of each animal as though it were a genealogical record, the meaning of this record has been neglected in favor of its illustrations. That was almost inevitable, given the vagaries in which embryologists have been enmeshed as a result of the overemphasis they place on human embryology.

Thanks to the many important studies of lower animals, we are now in a position to use morphology and comparative anatomy to trace the routes by which the higher organic types have developed. This will enable us to determine how the organisms belonging to each of these types have gradually become more complex in the normal course of embryonic development. We can foresee the possibility of determining exactly how each of these phenomena has been disturbed and discover the cause of these perturbations. The time seems to have come when it will be possible to fulfill Serres' hope and develop the close relationships of embryology to morphology and paleontology.

In this and the preceding chapter we have seen how we have arrived at our present notions as to what constitutes an individual organism. Relatively simple organic individuals have been formed from cells [elements anatomique] that are passed from one to another and can vary with the prevailing conditions or with their place in the genetic succession. Can these individuals, which are capable of propagating new forms, really be considered unchangeable when their own cells [éléments] or the groups of cells of which they themselves are composed are not permanent? This succession of individuals born from one another is precisely what we call a species. And this finally brings us to the question of the fixity or variability of species.

Chapter XX

Species and their Modifications

A brief review of ideas concerning species. – The true nature of the problem of species; means of resolving this problem directly. – Attempts to find solutions indirectly. – Differences between races and species. – Proposed criteria for defining species and their limited usefulness. – The instability of hybrid forms: – Godron's theory. – Charles Naudin's experiments and theory. – Identity of race and species. – Isadore Geoffroy Saint-Hilaire: the theory of limited variability. – Comparisons of the doctrines of Isadore Geoffroy Saint-Hilaire and Charles Darwin. – Conclusions.

Our interpretation of the physical structure of an individual species is closely linked to another question: were all the genealogical forbearers of a new species identical individuals that changed abruptly or did these individuals undergo gradual modifications that allow us to consider fossils from past geological periods as ancestors of the living forms and trace the species back to more and more simple forms that finally end with independent cells.

Linnaeus, Cuvier, de Blainville, Flourens, Dugès, and Louis Agassiz definitely decided in favor of the first of these alternatives. Partisans of the variability of species were equally numerous, but they interpreted variability in different ways. For Bonnet, the variability was only apparent; in the beginning, the germs were endowed with an organization appropriate to the diverse conditions of early geological epochs, and they developed when the prevailing conditions became favorable for them. Buffon believed that primitive species had been modified but that the modifications were the direct result of environmental effects and that they had simply degenerated from a well-established primitive predecessor. Etienne Geoffroy Saint-Hilaire, Goethe, and Richard Owen, assumed that beings have been created with their present degree of complexity and have been modified only in minor ways, much as Buffon had proposed but with stronger emphasis. Erasmus Darwin and Lamarck, on the other hand, thought that very simple forms created by God were born spontaneously and had gradually become more complex and better adapted before arriving at their present form. Which of these two opposed opinions was correct? Before Darwin published his classic book on the origin of species, various scientists had tried to find an answer by carefully considering all the facts known to science at that time, while a number of skillful experimentalists, such as Flourens, Kœlreuter, Godron, Isadore Geoffroy Saint-Hilaire, and Naudin, used other means to attack the problem. While their conclusions differed widely,

A. McBirney, S. Cook and G. Retallack, *The Philosophy of Zoology Before Darwin: A translated and annotated version of the French text by Edmond Perrier*,

their long discussions of the question of species led to all sorts of related questions, and instead of following the facts step by step in a scientific manner, they were carried away with the pros and cons of the different theoretical principles that had been proposed.

Take, for example, a pair of animals that are as closely related to each other as possible and consider the various individuals that could be born from their union. The new-born individuals, although siblings and necessarily of the same species, already have significant differences so that, when closely examined, they can always be distinguished from one another. It is obvious that certain characteristics of species can vary spontaneously. Consider next two sets of individuals born of the same mother and father. One set continues to live in the same conditions as that in which its parents lived, while the other is moved to a different climate and is placed in living conditions as far removed as possible from their parents. Surely, notable differences will appear between the two groups as they continue to develop. Under such conditions, if one allows the individuals making up each of the two groups to reproduce, it will generally come about that their differences will be accentuated with each successive generation and with time could become quite significant. Finally, if one returns the descendents of the separated groups to their original living conditions, their acquired characteristics will be preserved for a long time and will be transmitted almost intact to their descendents provided only the individuals with the same deviations from the primitive type are allowed to reproduce. The individuals with new hereditary characteristics will form a well-defined group within the species. They are what we call a *race*.

All species do not lend themselves equally well to forming races. There are some that, when transported into highly varied regions, will still preserve all their characteristics with a remarkable persistence. This is the case with certain cosmopolitan butterflies. While some of these species are not readily divided into races, it seems that others form new races with relative ease. This would be a good subject for research, but at this point, the only thing about which we can be confident is that they differ in their ability to vary and persist as different races.

Once races are established, they remain pure provided they reproduce only with others having all the same characteristics, and especially if these individuals continue to live in the same living conditions in which their race originated. Suppose now that a few individuals have produced a new race because their parents were moved into a region remote from where they originated. They will undergo modifications of their reproductive cells, including their sexual organs, the timing of their copulation, or even in their vital humors [humeurs de leur organisme] such that they will not be able to breed with individuals of the original location. The new and old races can live side by side without mixing, and according to all definitions except that of Agassiz, we call these races species. We are now able to propose the following hypothesis: individuals of the same species but of different races could undergo modifications of their reproductive apparatus or other parts of the organisms, and as a result of these changes they become completely isolated from the individuals that have remained identical to their common parents. This is the essence of the concept of species: the day when this separation is established scientifically the problem of species will be definitively resolved, even though a few particular cases may still pose difficulties. It is also the most direct way to resolve the question. Several proposals of this kind have been advanced, but unfortunately none is absolutely conclusive.

One might find a satisfactory solution to this problem by taking the opposite approach. Would it not be possible that closely related species which, when interbred, were always infertile, could become fertile if they lived together and had to adapt to the same living conditions? Several authors have thought that this must have been true of some of our domestic animals like goats, cattle, and especially dogs, many of which stem from a great variety of wild species that later began to mix. Here the argument misses an essential point, namely the proof that what were thought to be different species were really races of the same species. But what has been done in the past can be done again in the future, and the experiment is worth trying.

If both of these direct approaches to the problem prove inconclusive, one might try to get around the difficulties by studying the interbreeding of individuals that everyone considers distinct species, such as a dog and jackal, dog and wolf, or donkey and horse, camel and dromedary, sheep and goat, bull and doe, mountain goat and ewe, chamois and goat, the various species of llamas, hare and rabbit, the various species of fowls and birds, etc. One could try to find strict criteria for identifying species in this way and formulate relevant laws. Fréderic Cuvier said that only individuals of the same species can continue indefinitely to interbreed fruitfully; hybrids born from breeding individuals of different species are often sterile, and this sterility often appears after only a few generations. Flourens added that breeding between individuals of different genera is always sterile.

Frédéric Cuvier, Flourens, as well as Godron[1] were in agreement in considering the limited fertility of hybrids as a proof of the fixity of species. One wonders, however, whether the impossibility of creating permanent mixed breeds that are intermediate between two distinct forms proves that the present forms could not be modified to the degree that they would be incapable of interbreeding with those that still preserve the primitive characteristics of their common progenitors. These scientists seem to assume *à priori* that the species are fixed and instead of trying to find proofs of this fixity only offer arguments in its favor. Their reasoning and experimental methods might be very different if they were guided exclusively by facts and the conclusions one is justified in drawing from them.

Common every-day observations show that all sorts of beings perpetuate themselves in a limited number of forms that are always the same and have undergone no significant changes in all the time we have been observing them. These forms are what we call *species*. Our primary aim should be to explain this basic fact, and we will find the explanation in another fact, namely that animals and plants of different species are incapable of mixing and producing stable, intermediate forms, either because they cannot produce an offspring or, if they do, it is sterile.

---

[1] *De l'espèce et de la race chez les êtres organisés*, vol. 1, p. 217.

Physiologists continue to search for the causes of this infertility and sterility of hybrids, but they have yet to find an answer. Kœlreuter, Godron, and Charles Naudin have demonstrated that the reproductive cells of hybrids, especially those of the males, are defective, but the imperfections are irregular, and their cause has yet to be determined. That is as far as research has gone. Most authors have thought they could get around the problem by saying that the Creator wanted to maintain the purity of the species in this way, but this simply leads us into a vicious circle.

On the other hand, the barrier that the Creator is said to have established between two species is far from being uniformly effective. Hybrids seldom reproduce except with animals that are of the same or very similar genera. But within these limits, they present all possible degrees of fertility. Most often, only the males are infertile, and the females can be fertilized by the males of the parental species. This is the case for mules born to donkeys and mares. In other cases, as in that of the dog and wolf, the mixture can reproduce through several generations before the sterility appears. In yet another case, that of a hare and female rabbit, the mixed offspring remain fertile indefinitely, as though these animals that are generally so antithetic to one another are really the same species. Do these highly varied physiological characteristics of hybrids indicate that the distance separating one species from a closely related one is not always the same? The same is true if the neighboring species or even those that we consider to be of the same genus had descended from the same source. Experiments with hybridization, far from demonstrating the fixity of species, furnish arguments that species are formed by gradual modifications of pre-existing species. This is actually the conclusion to which Charles Naudin was led by his admirable research on interbreeding of numerous species of poppies, mirabilis, primula, datura, tobacco, gourds, etc.

Naudin remarked:[2] 'On contemplating the organized living world in which we live I was struck by the fact that, however varied their outward forms may be, organized beings share many important analogies. It is these analogies that enable us to classify them into *kingdoms, classes, families, genera, and species*. If one ignores them and assumes that each animal has its own independent regime, any logical classification becomes impossible. Is there any explanation for this remarkable phenomenon of analogies? Not if one holds to the belief that these forms are primordial and have remained unchanged. One must relate the varied forms of life to a system based on evolution from a common origin. For example, there are seven or eight hundred types of *Solanum* widely disseminated over the Old and the New World. Each has its distinct features, but they all share certain characteristics that, in the eyes of the taxonomist, are incomparably more important than the external differences that I consider superficial. It is their shared characteristics that enable us to assign them to a single class, family, and genus. Are these analogies simple, meaningless features with no physiological relationships, or do they exist simply because God wills it that way? If one believes in a system in

---

[2] Ch. Naudin, *Nouvelles recherches sur les hybrides végétaux Nouvelles archives du Muséum d'histoire Naturelle*. vol. 1, p. 169, 1865.

which each species has an independent origin, one must chose between pure chance (which is absurd) and a supernatural phenomenon, that is to say a miracle. There is no place in science for either of these. On the other hand, if one assumes that all the species have a common ancestor and applies this general rule to the plant kingdom where the present forms still preserve traces of their origin, it is possible to subdivide them into secondary forms that have diverged from a common origin. In doing this, one is forced to follow the basic principle of evolution and assign them to species, races, and varieties with slight differences. Superficial traits vary from one form to another, but the essential base they share is preserved. There could be a thousand derivative species, but each of them would carry the imprint of its origin, a sign of the parentage it shares with all the others. It is this sign that enables us to bring them all together as a single family and genus.'

Buffon was strongly opposed to this conclusion in the early years of his career when he saw naturalists using it in their classification systems, but he later accepted it and made it the basis of his own system.

Others have reached conclusions from studies of hybridization that are quite opposed to those of Naudin. They resort to other arguments to save the dogma of fixity of species by fashioning ingenious distinctions between wild and domestic species, between species and races, hybrids and cross breeds. Their philosophical system is well described in the following propositions laid out by Godron in his work, *De l'espèce et de la race chez les êtres organisés*:[3]

'1) The specific characteristics of wild animal species living today do not become modified, even under the influence of external agents. Because they remain unaltered, they provide a means for making clear distinctions between living animal species.

'2) The only modifications that they undergo are relatively modest; they arise accidentally and never become permanent, as long as the animal continues to live in the wild.

'3) Thus there are no natural races in the strict sense of the word. Races are the mark of human intervention.

'4) Species of wild animals that lived long before our time and have managed to survive down to the present geological period still preserve their basic forms. Their distinctive characteristics can be identified in remnants that have been preserved through the course of long centuries.[4]

'5) Despite the changes that could be produced by the effects of physical conditions to which the species were exposed, their basic organization has not been modified or transformed in any way that might cause one species to be confused with another or give rise to distinct new types. The animals living today are exactly the same as their progenitors that lived at the beginning of the present geological period.

---

[3] Godron, *De l'espèce et de la race chez les êtres organisés* vol. 1, p. 51.

[4] *Ibid.*, p. 144.

'6) Species have not varied any more during earlier geological periods than they have during our own. As a result, species living during those periods could not have been transformed into those that are our contemporaries.[5]

'7) If the supposed progressive transformation of beings were a reality and the simplest plants and animals had been perfected into more complex forms, or if the invertebrates had metamorphosed into vertebrates, fish, and reptiles and the reptiles into birds and mammals, or even the acotyledons into monocotyledons and then dicotyledons, such complete mutations would have required many centuries... In passing from one geological period to another, one would find intermediate forms in the course of transformation through all stages of these metamorphoses, and the animal kingdom, like the plant kingdom, would show a continuous series of forms varying so slightly that one could no longer find clear lines of demarcation between the characteristics of individual species. One would find nothing but confusion where we now see an admirable order. Far from that, we see instead abrupt interruptions between both the animal and vegetable forms of two successive geological periods. Distinct fauna and flora replace one another in a regular series of transformations. All these observations demonstrate the plurality and succession of special organic creations down through the ages of our planet.

'Species have varied no more during earlier geological time than they have during man's existence. One would expect the species that were originally created to display differences reflecting the effects of differing environmental conditions and the global revolutions for which we find indelible evidence in the stratigraphic record. These events could not have altered the types that were originally created. On the contrary, the species have remained stable until new conditions rendered their continued existence impossible. They then perished without having been modified.

'8) If species of wild animals do not vary and have remained unchanged since their creation, the same is not true of domesticated species. The latter have been subject for times as long as several centuries to highly varied living conditions and have undergone many important modifications of their physical characteristics, behavior, habits, and even their instincts. Domestication is such a strong modifying factor that its effects have been more complete and durable through longer periods of time.'[6]

Godron later added[7] that these modifications could become hereditary and, in this way, produce more durable races that would be clearly distinguished from species by the ability of members of different races of the same species to produce hybrids that remain fecund indefinitely. They transmit their mixed characteristics

---

[5] *Ibid.*, p. 332.

[6] Godron, *De l'espèce et de la race* vol. 1, p. 463.

[7] *Ibid.*, vol. 2, p. 46.

to their descendents and can generate unlimited numbers of intermediate races. He ends his theory of race with the proposition: 'If species are made by God, persistent races are made by man.'

Man himself is thought to be a unique species profoundly separated from the rest of the animal kingdom and deserves to constitute a unique realm that dominates the *moral realm* of Barbençois (1816) and *hominid realm* of Fabre d'Olivet (1822). Once this privileged being takes on attributes of the divine, we should not be surprised at anything that happens.

Thus, for Godron, when the species acquires its own identity it becomes perfectly immutable. The blind forces of nature are incapable of modifying it in any way. Created for a particular natural setting and living conditions, a species disappears when those conditions begin to change. With each global revolution, all creation is wiped out, and a new one marks the return to calm and stability. This creation remains the way God made it as long as the particular conditions for which it was instituted continue. In any case, the appearance of man opened a new era for the plant and animal species. Henceforth, an intelligence made in the image of the divine is going to shape the living forms to requirements that were previously unknown. These forms are going to cede to a certain degree to man's caprices, but the latter is not capable of creating new species – that privilege belongs only to God – humans produce only races and varieties.

It is impossible to erect a more complete system for this miraculous intervention into natural phenomena than the one we have just seen rejected so effectively by Naudin. But just as one cannot be a half-way evolutionist, one cannot be a half partisan of the fixity of species. All the predispositions [temperaments] that one finds in the two doctrines serve only to emphasize the discord, so often ignored, between factual evidence and the conclusions they inevitably entail. These cherished ideas have raised regrettable conflicts. The basic problem is that those who consider species immutable are forced to call upon miracles to support their beliefs. Those who believe in the theory of evolution see it as the essential basis of biological phenomena, just as, in the case of physical phenomena, the Creator acts entirely through conflicting forces.

Naudin was not mistaken. He credited human intelligence with no special powers. I was tempted to say that man has been delegated to play a special role with respect to the species. In his view, it was the environment that was responsible for everything:

'There is no qualitative difference,' he said, 'between *species, races*, and *varieties*, and it is pointless to pretend that there is. The three are really the same, and the words indicate nothing more than *degrees of contrast* between comparable forms… Their differences are of only a matter of degree, from the strongest to the weakest, which amounts to saying that when we attempt to compare them, we are only looking for degrees of strength and weakness, and our vocabulary is inadequate to express these degrees in words. Thus, as I have just said, delimitations of species are quite arbitrary. We make them broader or narrower according to the importance we assign to the resemblances and differences we see when the diverse groups are placed side by side, and the value we put on them varies according to the person, the times, and their scientific context.

'Does it follow that the words *race* and *variety* should be banished from science? Definitely not, for they are useful for designating forms that are not distinctive enough to be listed among established species. It is convenient to recognize their true importance, which is absolutely the same as that of proper species, and to see in the forms designated by these names weak indications of identity that can be neglected without serious scientific consequences.'[8]

Naudin considered species to be *a group of similar individuals that differ in some way from other groups and are propagated through a series of generations without losing the physiognomy and organization shared by all the individuals.*

At the same time, however, this capable botanist contributed to establishing a fact that future naturalists could invoke effectively in support of the fixity of species. From his studies of hybridization of plants belonging to the most varied groups, as well as numerous experiments on cross-breeding of animals, it turned out that the individuals that were direct products of these mixtures generally exhibit combinations of their parents' characteristics and can be considered almost exactly intermediate between them. However, if one allows these hybrids to interbreed for two or three generations, the characteristic differences tend to develop along separate lines, so that some individuals born of the same parents and belonging to the same generation closely resemble the father's species, while others resemble that of the mother. Intermediate types are rare and are usually very different from one another. In most cases, all the individuals end up reverting almost entirely to the species of one of the two parents, as if the blood of the other had been completely eliminated. Thus, under the conditions where they have been produced until now, the fecund cross-breeds cannot produce a species that is exactly intermediate between two others.

If, on the other hand, one crosses individuals that differ only in their race and continues to breed them exclusively with one another, the mixed progeny or *half-breeds* often preserve their intermediate characteristics through many generations. It is easy to create *mixed races* in this way, but it is impossible to create *mixed species.* This is what most eminent naturalists consider the distinctive attribute of races and species, and the distinction is certainly a valid one. No one would deny that there are groups of similar individuals that are isolated by their limited reproductive ability and that this renders formation of intermediate groups very difficult. Nothing prevents these groups from constituting species. Groups that share a common origin but are less isolated may be considered races, but there is a wide range of gradations between such races and true species. Some mixed races are less distinct and tend to disappear, and this allows one or both of the two parent

---

[8] Ch Naudin, *Nouvelles recherches sur l'hybridité dans les végétaux. Nouvelles archives de Muséum d'histoire naturelle,* 1st series vol. 1, 1865, p. 162). Although this memoir bears the date 1865, Charles Naudin had already expressed analogous ideas in 1852 in the *Revue horticole*, several years before the appearance of Charles Darwin's book on the origin of species.

races to be reestablished. The conditions in which the mixed breeds and hybrids are placed seems to have a notable influence on the durability of their characteristics.

As the eminent zoologist A. Sanson[9] says, this separation of the blood lines of two races that had been united in an intermediate race of mixed breeds and their later reversion to their two original types 'is not just an exception or even a rule; it is a law. This law has held for all known cases of reproduction between the individuals of two or more different races having their own fundamental characteristics or different specifics.[10] We can cite firm proof of this for all the animal genera that are subjects of animal husbandry [zootechnie].' Sanson found examples in the present state of all crossed races, including horses, cattle, sheep, pigs, dogs, and pigeons. Thus, the differences between hybrids and mixed breeds disappear, just as they do in the case of limited fecundity, and we must agree with Charles Naudin that the differences between races and species extend beyond the contrasting forms of close relatives. But if that were so, the doctrine of fixity of species would disappear completely. The specific forms enjoy a considerable degree of *stability*, but not true *fixity*. The *theory of limited variability* is based on this distinction between an acquired, revocable stability and an inherent, inalterable fixity. Isidore Geoffroy Saint-Hilaire devoted almost all of his *Histoire naturelle générale des règnes organiques* to demonstrating this principle.

Unfortunately, this fine book came out over a prolonged period from 1854 to 1862 and was never completed. It can therefore be considered a contemporary of Godron's book and the memoirs of Charles Naudin, and it was completely independent of the theories of Charles Darwin. Isidore Geoffroy discussed the variability of species and inter-breeding of domesticated animals using several lines of valid scientific evidence including the results of numerous experiments, many of which he carried out himself at the menagerie of the Museum of Natural History.

The conclusions of this long, scholarly study are summarized in the following propositions:[11]

'The characteristics of species are neither absolutely fixed, as several persons have said, nor completely variable, as others have maintained. They are fixed for each species and are perpetuated so long as the particular environment to which the species is adapted remains constant, but they can be modified if those conditions happen to change.

'In the latter case, the characteristics of the species are the *resultant* of two opposed forces: one is the *modifying* influence of the ambient conditions, the other the *conservative* hereditary effect that causes characteristics to be passed on from generation to generation.

---

[9] A. Sanson *Traité de zootechnie*, 2nd ed, vol. II, p. 62.

[10] Sanson uses the word *specific* here in the sense of zoological technology, which also applies to species of horses, cattle, sheep, dogs, as well as to solidly fixed races of these animals.

[11] Isidore Geoffroy Saint-Hilaire *Histoire [naturelle] générale des règnes organique*. vol. 2, p. 131, 1839. [This title seems to be confused with Histoire générale et particulaire des anomalies de l'organisation chez l'homme at les animaux 3 vol. 1832–1837]. (Trans. note)

'In order for the modifying effects to be much more dominant than the conservative hereditary tendencies a species must pass from the environmental conditions in which it lives into a new, very different situation where it always adapts to the ambient world. This explains the very narrow limits of observed variations in the case of wild animals. It also explains the extreme variability of the domestic animals.

'The species of wild animals generally remain in places and conditions where they are well established, and they spread as little as possible, because their organization is adapted to the conditions in these places and would not be in accord with other conditions. The same characteristics should therefore be transmitted from generation to generation.

'The conditions being permanent, the species are as well.

'Permanence and fixity are not absolute, however. In the long term, the gradual expansion of the species over the surface of the globe is a necessary consequence of the multiplication of individuals. Other causes of a less general nature can also lead to minor displacements.

'This means that, within the limits of the geographical distribution of the most wide-spread species, notable differences of habitat and climate can bring about second order differences in their regime and even in their behavior. These various kinds of differences are seen in races characterized by different colors and other external characteristics, such as the proportions and size of their bodies, and in some cases their internal organization. These races have sometimes been arbitrarily called local varieties, and some have even been taken for distinct species.

'Among domesticated animals, the causes of variation are much more numerous and stronger. In a long series of experiments that were undertaken for a quite practical purpose and with no theoretical basis, species of several classes numbering about forty in all, were taken from their wild state and allowed to adapt their behavior to very diverse regimes and climates. The effects obtained can be directly attributed to known causes. The result was to produce a multitude of very distinct races, several of which even have characteristics every bit as strong as those that ordinarily distinguish genera.

'A number of domesticated races have returned to the wild state in diverse places throughout the world, and this has provided a second series of experiments that are the inverse of the preceding ones and provide evidence for the counter effect. If domesticated animals are returned to the conditions in which their ancestors lived, it takes only a few generations for the descendents to recover their original characteristics, but they do this only if they are put back into a wild setting with analogous though not necessarily identical conditions...'

Isidore Geoffroy Saint-Hilaire, in contrast to Godron, has presented arguments that are hard to refute when he showed that the limited variability of species has been fully demonstrated by both observations and experiments. He added, moreover, that the principles he explored 'can lead to rational solutions to questions for which the partisans of absolute fixity have no answers whatever or have attempted to resolve by appealing to the most complex and unreasonable hypotheses.

'This has become a fundamental issue for anthropologists. The common origin of all the diverse human races is rationally admissible from the point of view of their variability and from that alone. When the partisans of absolute fixity agree with us on this, they are forced to abandon most of their own long-held principles.

'In paleontology, only one simple, rational hypothesis is consistent with the theory of limited variability, that of genetic relationship [filiation]; the doctrine of fixity rests on two equally complex and unreasonable hypotheses, that of *successive creations* and what they call *translation*.'

Isidore Geoffroy naturally sided with the hypothesis of genealogical relationship that allows us, for example, 'to search for the ancestors of our elephants, rhinoceros, and crocodiles among the elephants, rhinoceros, and crocodiles that paleontology has shown existed in antediluvian times.'

Meanwhile in England, Darwin was giving the doctrine of evolution a vitality it never had before, while in France the illustrious heir to the name Geoffroy was becoming a steadfast defender of this same [Darwinian] doctrine. Without the slightest doubt, if death had not overtaken him at the moment when science could still gain much from his laborious, patient, and impartial investigations, Isidore Geoffroy would have enlarged the basis of his theory, and this would have made him a kind of compromise between the two scholars who represented the two sides of the narrow gap between analogous ideas. But we can take the theory of limited variability only to the point where Geoffroy led it, and we must specify in what way it differs from the doctrine of Charles Darwin.

First of all, what meaning did the adjective *limited* lend to the word *variability*? Are limits imposed on the extent of the variations that a specific form can undergo, or should these limits apply during the time when these variations could be effected, so that variability would be *limited* to certain epochs? It seems probable that both of these two interpretations had an equal status in the mind of Isidore Geoffroy. When one looks at the surface of the Earth, the average living conditions offered to living beings in the wide variations of their surroundings seem, first of all, to oscillate within rather narrow limits. These limits determine the modifications that species can undergo and are always narrowly dependent on external agents. The greatest variations of conditions have probably been mainly during intervals separating one geological period from another, but the intervening epochs might also have been capable of bringing about great transformations of species.

Nowhere in his work did Isidore Geoffroy say how far he would be willing to extend his interpretation of these transformations, but as soon as one accepts the hypothesis of evolution it becomes impossible to limit the manner in which it might be applied. It now seems well established that during the Paleozoic era there were neither birds nor mammals, that the reptiles appeared only after the amphibians and fish, and that the fish themselves appeared only after the invertebrate animals. The order of succession of the mammals during the Tertiary period could have been arranged in a very remarkable manner. A strict application of the general concept of descent with modification implies that animals have been drawn from one another and obviously could not have been modified in this way without also attributing to species a variability that is regulated by precise but undefined laws. If the variations that a species can undergo during a geological period appear to be limited, it is impossible to believe such a restriction could hold through the entire duration of time.

But can one accept the premise that, during a long geological period, species remain so stable that they are unable to form anything more than regional races? Such a hypothesis is obviously tied to the assumption that throughout the history of the world there have been alternating periods of rapid change and stasis. Geologists seem to be distancing themselves more and more from this view. There is increasing evidence that the surface of the earth is always changing at the same slow pace we see today, and that there have been no sharp demarcations between successive geological periods. One must therefore accept the fact that species are always capable of changing and that the words 'limited variability' no longer mean anything more than slow but unlimited variability while adapting to ambient conditions according to the laws of heredity.

In order for this variability to function, is it necessary, as Isidore Geoffroy proposed, that there be important modifications of the Earth's ecological conditions? Certainly not. He himself pointed out that the gradual spread of species over the globe is a necessary consequence of the growing populations that change the conditions under which these individuals must live and makes them susceptible to modifications. But what limit can be placed on the effect of this expansive force? Could it not eventually reach a stage when long successions of generations are able to adapt to very different conditions? Is it necessary to assume changes in an environment that is already quite varied, if the individuals of a given species are forced, under pain of death, to adapt themselves to the most diversified kinds of life and spontaneously go in search of the most varied types of environments? Obviously not. It is this that Charles Darwin has so brilliantly demonstrated, and it is in this respect that his doctrine, though incomplete, differs from that of Isidore Geoffroy Saint-Hilaire.

The French naturalist might say that the organisms are transformed passively in response to environmental changes to which they are subjected. The English naturalist would say that the active multiplication of individuals and the struggle for life that results from it, obliges the animals and plants to exploit all the conditions of existence that are available to them. The infinite variety of the environment can remain unchanged, but the species is malleable and capable of taking on varied forms in order to continue its unlimited expansion. If this is the case, the range of possible modifications is limitless. On the one hand, members of the same species permanently preserve some distinctive trace of their common origin, and on the other hand, the progeny of each species has the possibility to grow and spread into one of the innumerable domains that the Earth offers for the expansion of fecund species. Isidore Geoffroy showed us modifying agents that function in some sort of intermittent way; Charles Darwin showed us, in addition to these agents, a more important modifying force of infinite power that determines how these agents become effective. It is the expansive force that results from the ability of individual members of a species to reproduce. According to this new hypothesis, the species have never ceased being modified since the time when life first appeared on Earth, and one can easily understand how living forms have reached the prodigious diversity that studies of botany, zoology, and paleontology have revealed to us. To explain the modifications to which the species are susceptible it is no longer necessary to appeal to extraordinary unknown processes that have never been witnessed by man. It is not even necessary to postulate more or less profound changes in the environment in which the organisms live. The modifications of living forms are, like all the physical and chemical phenomena we observe, the result of tangible causes that are much more effective.

We now realize that the problems of zoology and botany must be posed quite differently from the way they were seen by earlier naturalists. Each living form has appeared as the result of a series of actions of the environment on its ancestors, and one recognizes the possibility of determining which of these actions produced them and in what order they have followed one another.

It is no longer a matter of setting up a simple portrait of Nature; it is no longer a question of revealing mysterious motivations. It is not even a question of defining the laws that govern the production of organisms. Instead, we must find true explanations for the development of each living being, and these must be explanations in the sense that physicists and chemists understand the word. To do this, the natural sciences must adopt methods of the kind normally used in the physical sciences. The true superiority of the doctrine of evolution is in the methods that have been used in developing it, and now that Darwin has made a start on this, we are witnessing a true renaissance in all branches of natural history. It is true that we are still far from having reached the brilliant results we dream of achieving, but is it not more important that we have detached ourselves from the narrow anthropomorphism that for so many centuries has burdened the thinking of naturalists? We now understand that the explanations of living beings will be found in the world that they inhabit and not in external causes. We are now convinced that there will be no progress in biology until the day when one can identify the origins of every organic form and explain the way it was produced. A zoological classification will be nothing more than a history of the successive adaptations through which life has evolved.

Naturalists have long thought that this goal was beyond their reach. Until the first half of this century, they were tired of searching nature for elusive explanations and believed they had to attribute each living form to the intervention of a supernatural will. I hope that I have demonstrated in the preceding pages that their new endeavors are fully justified by the results already obtained, even though this may have raised difficulties of another kind. The former doctrine which viewed nature as the work of an all-powerful creator seemed to put man in constant collaboration with God. It was feared that, in portraying life as having risen like inanimate bodies through the blind action of the physical forces, there would be no place left for the Creator. But that view is just another form of anthropomorphism.

To those who may be tormented by such scruples, it is helpful to recall that chemistry, physics, and astronomy, in explaining the factual evidence on which their respective domains are founded, have yet to discern a first cause. In this respect, modern biology is no different. It does not suppress God; it sees Him as more remote and, above all, at a higher level.

END

References

Agassiz L, 1857, Contributions to the natural history of the United States, Essay on classification. London;

_________, 1862, *De l'espèce et de la classification en zoologies.* Paris

Appel T A, 1987, *The Cuvier – Geoffroy Debate: French Biology in the Decades Before Darwin.* Oxford University Press

Bichat X, 1831, *Anatomie générale.* Blandin, Paris

Blanchard E, 1846, *Recherches anatomiques et zoologiques sur le systeme nerveux des animaux sans vertebras Annales des sciences naturelles*, 3rd series

Bonnet Charles, 1770, *La Palingénésis Philosophique, ou Idées sur l'État passé et sur l'État future des Êtres Vivants.* 2 vol, Geneva

_________, 1779, *Considérations sur les corps organisés (Oeuvres d'histoire naturelle et de philosophie.* Fauche, Paris

Bowler Peter J, 2003, *Evolution, The History of an Idea.* 3rd Ed, Univ of California Press, Berkeley

Buffon George-Louis Leclerc, Comte de, 1749–67, *Histoire Naturelle , Générale et Particuliere.* 15 vols, Paris

Butler S, 1882, *Evolution, Old and New: Or the theories of Buffon, Dr. Erasmus Darwin, and Lamarck, as Compared with that of Mr. Charles Darwin and Lamarck.* 2nd Ed. London. Reprinted 2008 Bibliobazaar, Charleston

Cahn T, 1962, *La vie et œuvre d'Etienne Geoffroy Saint-Hilaire.* Presses Universitaires de France, Paris

Coleman William, 1964, *George Cuvier, Zoologist: A Study in the History of Evolution Theory.* Harvard Univ Press, Cambridge Mass.

_________, 1971, *Biology of the Nineteenth Century: Problems of Form, Function, and Transmutation.* Wiley, New York

Cuvier G, 1826, *Discours sur les revolutions de la surface du globe, et sur les changemens qu'elles ont produit dans le règne animal.* 2d Ed, Didot, Paris, Reprint, 1969, Culture et Civilization, Brussels

_________, 1831, *Mémoire sur l'oreille osseuse des crocodiles et des téléosaures*

Darwin, Charles, 1858, *Variations of animals and plants under domestication.* 2 vols, London. Reprint, 1998, Johns Hopkins University Press, Baltimore

_________, 1859, *The Origins of Species by Means of Natural Selection: Or the Preservation if Favored Races in the Struggle for Life.* London. Reprint, 1964, Harvard University Press, Cambridge, Mass.

_________, 1871, *The Descent of Man and Selection in Relation to Sex.* 2 vols, London. Reprint, 1972, AMS Press, New York

Darwin Erasmus, 1794–96, *Zoonomia: Or the Laws of Organic Life.* 2 vols, London. Reprint, 1974, AMS Press, New York

Diderot, D, 1754, *Pensées sur L'interpretation de la Nature.* Paris

Dodson E O and Dodson P, 1985, *Evolution: Process and Product*, 3rd Edit. Prindle, Weber & Schmidt, Boston

Dugès A, 1828, Recherches sur l'organisation et les mœurs des Planaires. *Annales des sciences naturelles*, 1st series, vol. XV

_________, 1832, *Mémoire sur la conformité organique dans l'échelle animale.* Montpelier

_________, 1871, *Les origins animals de l'homme*, 8 vol, Germer Bailliere, Paris

_________, 1850, Aperçu de quelques observations nouvelles sur les Planaires et plusieurs genres voisins. *Annales des sciences naturelles* 1st series

Friedenthal, Richard, 1963, *Goethe: His Life and Work* Cleveland World Publishing Company, Cleveland

Geoffroy Saint-Hilaire, Etienne, 1818, *Philosophie anatomique: Les organs réspiratoires sous le rapport de la determination et de l'identité de leur pieces osseuses.* 2 vol, Doublet, Paris. Reprint, 1968, Culture et Civilization, Brussels

_________, 1835, Études progressives d'un naturaliste pendant les années 1834 et 1835. *Muséum d'Histoire Naturelle*, Paris.

Geoffroy Saint-Hilaire, Isidore, 1841, *Essaies de zoologie general, ou, Mémoires et notices sur la zoologie générale , l'anthropologie, et l'histoire de la science.* Paris

_________, 1854–1862, *Histoire Naturelle Générale des Règnes Organiques.* 3 vol, Paris.

_________, 1847, *Vie, travail et doctrine d'Etienne Geoffroy Saint-Hilaire.* Reprint, 1968, Culture et Civilization, Brussels

Gillispie, C C, 1959, Lamarck and Darwin in the History of Science, in *Forerunners of Darwin: 1745 – 1859*, B Glass, ed. Johns Hopkins Press, Baltimore.

_________, 1980, *Science and Polity in France: The End of the Old Regime,* Princeton University Press, Princeton NJ

_________, 2004, *Science and Polity in France: The Revolution and Napoleonic Years*, Princeton University Press, Princeton NJ

Glass, B, Temkin, O, and Straus W. L., eds.. 1959, *Forerunners of Darwin: 1745 – 1859.* Johns Hopkins Press, Baltimore

Godron, Dominique Alexandre, 1848, De l'Espèce et des Races dans les Etres Organizés et Spécialment de L'Unité de l'Espèce Humaine, *Mémoire de la Societé Royale des Science de Nancy*

Haeckel, E, 1882, *Die Naturanschauung von Darwin, Goethe, und Lamarck.* Jena.

Hyman Libbie H, 1940, *The Invertebrates: Protozoa through Ctenophora.* McGraw-Hill Book Co, New York

Lacaze-Duthiers H, 1829 and following years, Recherches sur l'armure génitale femelle des insects *Annales des Science Naturelles*, 3rd series

Lamarck J-B P A de, 1802, *Recherches sur l'Organization des Corps Vivants.* Paris

_________, 1873, *Philosophie Zoologique.* 2 vol, Charles Martin, Paris.

_________, 1835-1845, *Histoire Naturelle des Animaux sans Vertèbres* 2nd Ed, 6 vol, Paris. Reprint, 1960, Culture and Civilization, Brussels

Le Guyader H, 2004, *Geoffroy Saint-Hilaire: A Visionary Naturalist.* An English translation by Marjorie Grene of the original1998 edition, Univ. of Chicago Press, Chicago

Leuckart C A, 1851, *Ueber den Polymorphismus der Individuen oder dir Ercheinungen der Arbeitstheilung in der Natur.* Giessen

Linnaeus, Carolus (Karl von Linné), 1735, *SystemaNaturae* Reprint, 1964, edited by M. S. J. Engel-Ledeboer and H. Engel, Nieuwkoop: B. de Graff

Lovejoy A O, 1959, Buffon and the Problem of Species, in *Forerunners of Darwin: 1745 – 1859*, B Glass, ed. Johns Hopkins Press, Baltimore.

_________, 1959, *The Argument of Organic Evolution before the Origin of Species*, in *Fore-runners of Darwin: 1745 – 1859*, B Glass, ed. Johns Hopkins Press, Baltimore.

Loven S, 1841, Observations sur le développement et les metamorphoses des genres Campanu-laires et Syncoryne. *Annales des sciences naturelles* 2nd Series, v. XIV

Lurie, Edward, 1960, *Louis Agassiz: A Life in Science,* University of Chicago Press, Chicago.

Maillet, M. de, 1754, *Telliamed ou Entretiens d'un Philosophe Indien sur la diminution de la Mer, avec un Missionaire Français.* 2 v, Paris

Maupertuis, P. L. M. de, 1751, *Systeme de la Nature*, Paris

________, 1768, *Oeuvres* 4 vol, Reprint, 1968, Georg Olms, Hildesheim

Milne-Edwards H, 1834–1840, *Histoire naturelle des crustacés, comprenant l'anatomie, la physiologie et la classification de ces animaux.* 3 vol, Paris

________, 1844, Considerations sur quelques principes relatifs à la classification naturelle des animaux. *Annales des Sciences Naturelles*, 3rd series vol 1, p 65

________, 1853, *Introduction à la zoologie général.* V Masson, Paris

________, 1870, Sur les travaux de Ch. Darwin, *La Revue des Cours Scientifique* 7: 591

Osborn H F, 1922, *From the Greeks to Darwin: An Outline of the Development of the Evolution Idea.* Macmillan, London

Owen, Richard, 1846, *Lectures on the Comparative anatomy and physiology of the vertebrate animals, delivered at the Royal College of Surgeons of England in 1844 and 1846.* London.

________, 1848. On the archetype and homologies of the vertebrate skeleton. London. Reprint AMS Press, New York

________, 1855, *Principles of Comparative Osteology*, London

Paley, William, 1802. *Natural Theology or Evidences of the Existence and Attributes of the Deity collected from the Appearance of Nature.* London. Reprint, 1970, Gregg International, London

Pasteur, Louis, 1864, Chimie appliquée à la physiologie: Des generations spontané, *Revue des Cours Scientiques* I: 257.

Perrier E, 1881, *Les colonies animals et la formation des organisms.* Paris

Quatrefages J-L A de, 1862, *Physiologie comparée: Métamorphoses de l'homme et des animaux metamorphoses de l'homme et des animaux.* Baillière, Paris

________, 1869, Histoire naturelle générale: Origines des espèces animals et végétales: Discussion des theories transformistes; L'Espèce et la race. *Revue des Deux Mondes*, 8th ser.

________, 1892, Darwin et ses Précurseurs Français, Étude sur le Transformisme. *Bibl. Scient. Internat*, LXXV, Paris.

Rathke M H, 1829, *Ueber die Bildung und Entwicklung des Flusskrebses*, Leipzig.

Robinet, J-B, 1768, *Considerations Philosophiques de la Gradation Naturelle des Formes de l'être, ou les Essais de la Nature qui apprend à faire l'Homme,* Paris

________, 1766, *De la nature.* 4 v, Amsterdam

Rudwick, M. J. S., 1997, *Georges Cuvier, Fossil Bones, and Geological Catastrophes*, University of Chicago Press, Chicago

Schelling, Friedrich Wilhem Joseph von, 1833–4, *Zur Geschichte der neueren Philosophie.* On the History of Modern Philosophy (1994) translation and introduction by A. Bowie, Cambridge University Press, Cambridge

Steenstrup J, 1842, *Ueber den Generationsvechsel, oder die Fortpflanzung und Entwickelung durch abwechselneden Generationen, eine eigenthumlehre Form der Brutpflege in den niederen Thierclassen.* Copenhagen

Sulloway Frank J, 1982. Darwin and His Finches: The Evolution of a Legend. *J Hist. Biol* 15: 1–53.

________, 1982, Darwin's Conversion: The Beagle Voyage and Its Aftermath. *J Hist Biol* 15: 325–96.

Swammerdamm J, 1752, *Biblia naturæ*

Teilhard de Chardin P, 1959, *The Phenomenon of Man.* Translations from the French by J. Huxley of Le Phénomène Humain, 1955

Thomson, Keith, 2005, *Before Darwin: Reconciling God and Nature,* Yale University Press, New Haven and London.

Van Beneden P-J, 1847, Mémoire sur les cestoïdes. *Bull Acad Bruxelles*

Von Baer K E, 1828, *Über Entwickelungsgeschichte der Thiere: Beobachtung und Reflexion.* Konigsburg. Reprint, 1967,Culture et Civilization, Brussels

Glossary

Certain terms present special problems, because their meanings may be obscure (cf. *emboitement*) and have changed with time (cf. *final cause*). Some are "false friends" in that the meaning of a word may differ between French and English. Many are so technical and rare that we have had difficulty in finding their definitions and translations (cf. *méride)*; and our English vocabulary is often imprecise (cf. *needs, in a Lamarckian sense rather than wants in the sense of desires*).

Acalèphe, *n.*, Medusa. "The general name given by the naturalists to non-sedentary polyps, such as the medusas, physalias, and beroës (Nom général donné par les naturalistes aux polypes non sédentaires, comme les méduses, les physalies, les béroés). (Émile Littré, Dictionnaire de la langue française (1872–1877))." Cuvier mentioned several genera of ctenophores in his class Acalephes of the higher taxon, Zoophyta (Hyman, 1940, pp 27–28).

Agamic, *adj.,* asexual reproduction, from the Greek agamos for unmarried.

Agamous, *adj.*, a variant of agamic.

Allantois, *n.*, a vascular fetal membrane of reptiles, birds, and mammals that is formed as a pouch from the hindgut and that in placental mammals is intimately associated with the outer embryonic membrane in forming the placenta.

Anatomical element, *n.*, Perrier frequently refers to "éléments anatomiques." The meaning of the phrase depends on the context and date of its use, its intention usually being "fundamental unit" or "building block." Early on, it was used to refer to organs. For instance, Henri Milne-Edwards (1853. Introduction à la zoologie général. Paris. V. Masson.) referred to muscle fibers, body segments, and lateral appendages of invertebrates and vertebrates with this term. But twenty years later, Charles Robin (1873. Anatomie et physiologie cellulaires; ou, des cellules animales et végétales.... Paris. J. b. Bailliere et Fils. Page 1) would say, "La plupart des individualités élémentaires organiques débutent par cet état dit de cellule;" However, he recognized that, although organisms are composed primarily of cells, not all "éléments anatomique" are cellular. Take, for instance, structural material secreted by or caused to be deposited by cells.

Perrier makes clear to us the interrelations among the words, anatomical element, sarcode, protoplasm, plastid and cell: "the anatomical element can be defined only as *a mass of protoplasm or sarcode of limited size endowed with independent life, ordinarily with a nucleus in its interior and capable of isolating itself by means of a more or less resistant membrane*. This anatomical element [cell] comprises what Hæckel called a *plastid*, a simple term that we can now adopt, even though it is relatively recent [Perrier, p 188]." We have frequently, but not universally, translated "élément" or "élément anatomique" as "cell."

Annelid, *n*., a common name for about 9000 species of wormlike invertebrate animals with well-developed segmentation.

Ascaride, *n*., any of the family of nematode worms that include the common round-worm that is parasitic in the human intestine.

Asexual reproduction Reproduction or propagation by mitotic division of cells, unrelated to the sexual (meiotic) processes of formation of gametes by diploid organisms or fusion of gametes to create the diploid state. Cf. budding (gemmation) or division (fission). According to Perrier (Ch 17 pp 178 and 179), Richard Owen gave the word *metagenesis* to *asexual reproduction.* E.g. Perrier, Ch 17, p 179: "The phenomena of reproduction that are most general and that do not require fertilization can be designated by the name chosen by Richard Owen, *metagenesis*."

Blastoderm, *n*., the modern definition of the term (Oxford English Dictionary): "The germinal skin or membrane surrounding the yolk in the impregnated living ovum, and constituting the superficial layer of the embryo in its earliest condition. It divides into two and afterwards three layers of cells ... from one or another of which all the parts of the new animal are developed." In 1859, it was more simply defined: "A layer of nucleated organised cells, named by Pander Blastoderm or germinal membrane."

Brachyura, *n.,* true crabs.

Cause, *n*., final, ultimate; first or primary (premiere). One of Perrier's main concerns was to place biology on the empirical and analytical basis of the physics and chemistry of his time. Consequently, he is at odds with the advocates of natural theology, who would find support for the efficacy of a Creator in the properties of organisms. The doctrine of final causes is essentially that of teleology. Indeed these terms are equated in the definition of teleology provided by the Oxford English Dictionary (Second Edition, 1989): "the doctrine or study of ends or final causes, esp. as related to the evidences of design or purpose in nature; ...."

Perrier sought a return to what later would be considered Aristotle's original intention with his fourth cause (telos), namely that of the functional significance of the end- or adult-state of a developed object, be it natural or human artifact. The above-mentioned fourth definition of cause of the Oxford English

Dictionary reflects the "Christianization" ("Scholasticism", Encyclopedia Brittanica, 14th ed. Vol 20, pp 81–82; no author) of Aristotle's concept by the scholastics of the middle ages: "The object of action; purpose, end. Obs. exc. in … final cause: a term introduced into philosophical language by the schoolmen as a transl. of Aristotle's fourth cause, or the end or purpose for which a thing is done, viewed as the cause of the act; esp. applied in Natural Theology to the design, purpose, or end of the arrangements of the universe." Perrier might have read the following definition in the Dictionnaire de L'Académie française, 6th Edition (1832-5): "CAUSE. s. f. (Page 1:271) Cause finale, Ce qu'on se propose pour but. La gloire de Dieu doit être la cause finale de toutes nos actions. Il se dit plus particulièrement de la Fin, du but pour lequel on suppose que chaque chose a été faite, créée. La doctrine des causes finales."

The definition of cause in the first edition of the Dictionnaire de L'Académie française (1694) enables us to draw a clear distinction between Aristotle's four causes and the concept of premier (first) or supreme cause: "Cause. Feminine noun. In principle, that which causes something to be. God is the first (premier) of all the causes, the cause of causes, the sovereign cause. One calls God absolutely and unexceptionally the premier Cause, as one calls creatures secondary causes. God effects the second causes. There are also different sorts of causes: principal, instrumental, material, formal, efficient, final, exemplary, physical, and moral." Aristotle's four causes are material, formal, efficient and final. Hence, Aristotle's first cause is definitely not the premier or "first" cause. For an example of Perrier's usage see the penultimate sentence of his book where he refers to God as the "cause première."

Cell, *n.*, Perrier refers to the cell of organisms by the following terms: cellule, plastide, and élément. By element, he conveys the idea that they are the elementary building units of a tissue or body, much as the atomic elements were thought to be the elementary units or atoms of matter. For example, Perrier spoke of anatomical and histological elements. Cells of multicellular organisms are homologous to the cells of infusoria. This latter inference is made clear in Perrier's formulation of one of Oken's à priori assumptions: "Les animaux et les plantes sont composés d'élément originairement semblables entre eux, analogues à des infusoires, les cellules. [p 167 in Ch XIV of original text]." On one occasion, Perrier refers to a cell (spermatozoid) as a "filament mobile:" "La division de l'oeuf peut être ou non précèdée de son mélange intime avec un élément semblable à lui ou en forme de filament mobile. [p 251 in Ch XVIII of the original text]."

Cestode, *n.*, a parasitic flat worm such as the tapeworm.

Cetaceans, *n.*, marine mammals, such as whales and porpoises.

Chiton, *n.*, a marine mollusc of the order polyplacophora with elongated bilateral symmetry and a dorsal shell of calcareous plates.

Cirrhipeds, *n.*, barnacles.

Coleoptera, *n.*, beetles.

Cryptogam, *n.*, a plant, such as a fern, moss, alga, or fungus, that reproduces by spores rather than by seeds.

Cypris, *n.*, 1) a small, fresh-water ostracod crustacean with long antenna that can be used for walking. 2) the second larval stage of barnacles.

Dasyure, *n.*, a kind of carnivorous, nocturnal marsupial found in Australia and New Guinea; also known as a "native cat". It has a long bushy tail and a dark coat with white spots.

Deme, *n.*, "Any undifferentiated aggregate of cells, plastids, or monads. (Applied by Perrier to the tertiary or higher individual resulting from the aggregate integration of mérides or permanent colonies of cells.)" Meaning 2 of the Oxford English Dictionary, 2nd Ed. 1989. The word also has other quite different meanings: judge, arbiter, or ruler (obsolete); a township in ancient Attica, and, in modern Greece, a commune; and "A local population of closely related plants or animals."

Dentalium, *n.*, a marine mollusc with the shape of a trumpet that lives in sand or mud.

Diptera, *n.*, derived from the Greek words "di" meaning two and "ptera" meaning wings, refers to true flies having only a single pair of wings.

Emboitement, *n.*, The theory of *emboitement,* propounded by Bonnet, is that of *preformation* (q.v.) in English. Perrier refers to it also in two other ways: as the hypothesis of invisible germs and the hypothesis of pre-existing germ cells. The word derives from "boîte," "box," and applied originally to a joint or joining of two pieces, one within the other, as in a joint of bones or the "emboîtement" of mortices and tenons in carpentry. Bonnet reasoned that within each original organism were germs within germs (invisible) that served as a template for succeeding generations – where by *germ* he meant a completely preformed organism. (See *emboitement* in http://dictionary.sensagent.com or in the Oxford English Dictionary).

Epigenesis, *n.*, "The concept according to which the body develops gradually by the addition of its diverse parts; the new being passing in stages from egg to germ [zygote] to embryo [and so forth]." From: Émile Littré, Dictionnaire de la langue française (1872–1877). Epigenesis is defined, more or less, by Perrier in Chapter 5.

Evolution, *n.*, The Oxford English Dictionary is an excellent source for understanding the many meanings and history of usage of this word. For our purposes, it has been used in biology primarily in two senses: 1) development or ontogeny of the individual from germ or embryo to adult or in a more general sense, 2) descent with modification or phylogenesis (as used by Lamarck, Darwin and Wallace).

We can illustrate the pitfall presented to a translator by citing the following passage, in which Perrier discusses Bonnet's concept of development or ontogeny (Ch 6, p 37): "In other words, there is no such thing as *generation* in the sense of producing a new form of life from a progenitor, only *evolution* [thus, *development*] of a pre-existing germ."

Gemmation, *n.*, asexual reproduction in which the new body becomes detached from a parent plant. (See: Asexual reproduction).

Geneagenesis, *n.*, According to Perrier (Ch 17 p 180 of our text), this is the name given by de Quatrefages to the production of large numbers of individuals by asexual reproduction. In Perrier's words, "the egg does not engender a single organism, but a whole [multitudinous] generation." The Oxford English Dictionary (Second Ed. 1989) defines it differently as "alternation of generation." This meaning is from H. Lawson's 1864 translation of Armand de Quatrefages', Physiologie comparée: metamorphoses de l'homme et des animaux. Paris, Baillière, 1862.

Germe, *n.*, Bonnet imagined the existence of invisible and indestructible embryos (germes) in both plants and animals produced by the Creator (see Ch 6). He reasoned that these preformed germs were enclosed (or boxed) one within another (cf. emboitement). Germe takes on other meanings in other contexts, for instance as in "germ cell" which refers to gametes, wheat germ, and the germinal disk of a bird's egg.

Helminthe, *n.*, a parasitic worm.

Homology, *n.*, See discussion under "Unity of plan of organic composition."

Hydra, *n.*, a marine or fresh water cnidairian invertebrate that absorbs CO_2 and uses sunlight for photosynthesis.

Integument See Tegument.

Lamarckism, *n.*, Lamarckism and its basic premises are not fully defined by Perrier. He does not mention the fundamental differences from Darwinian principles that were at the center of much of the debate. Lamarck assigned greater importance to harmonious adaptation to a changing environment rather than to conflict and competition for limited resources. Organs developed or atrophied according to the use entailed by the animal's effort to meet needs. Lamarck's work was presented in detailed scientific publications that, unlike *Origin of Species*, had little popular appeal and have been read by few of his critics. As a result, his ideas have often been misrepresented.

Lamellibranch, *n.*, any of a class of bivalve molluscs, such as clams, oysters, and mussels, that have a bilaterally compressed body enclosed within its mantle and a shell whose right and left parts are connected by a hinge over the animals back.

Law of continuity "Principle of Gottfried Wilhelm Leibniz (1646-1716) which can be roughly rendered as saying that when the difference between two causes is diminished indefinitely, so too is the difference between their effects (though Leibniz would not put it in these causal terms, since for him God is the only true cause). 'Nature makes no leaps', as he says in the Preface to his *New Essays on the Human Understanding* (c. 1704)" (http://www.philosophyprofessor.com). Perrier also refers to Aristotle's concept of continuity.

Maki, *n.*, a primate of the lemur genus with a long snout and long tail. It is found exclusively in Madagascar.

Medusa, *n.*, a gelatinous marine animal with tentacles suspended from an umbrella-like body; jelly fish. (See Acalèphe).

Méride, protoméride, *n.*, "In the morphological theory of French zoologist E. Perrier: [méride] a group of cells forming a distinct and permanent unit, either living as an independent organism or forming part of a colonial aggregate (Oxford English Dictionary)." "Protoméride (zoologie). – Premier individu (méride) d'une colonie animale. Dans les associations linéaires , le protoméride constitue la tête." (Dictionnaire Imago mundi, at http://www.cosmovisions.com/). As used by Perrier, the terms have an additional meaning when he discussed the concept of levels of biological organization. The first order of organization is that of the cell; the second is that of the *méride*. He defines méride as follows (p 196 of the present text): "Fertilization almost always precedes division of the egg when the daughter cells remain associated; absence of fertilization constitutes parthenogenesis. The cells that remain associated are not constrained to keep a single form; after they differentiate they form *a relatively simple organism of undefined type* ("une organisme relativement simple, sans type défini,") to which we shall give the name méride ("auquel nous donnerons le nom de méride"). *So, the méride is a simple undefined mass of cells.* But, more than that, it is the archetypical multicellular primordium that serves as a building block for the third order of organization. (We may compare this hypothetical body to Goethe's "Urblatt," for, as we shall see, the méride is equivalent to both the archetypical zoonite = somite = metamere; and the archetypical zooid).

In Perrier's schema, the méride may replicate itself, and the resulting entities may separate as ramets (bringing about geneagesis): "Les mérides se multiplient comme les plastides: tantôt ils produisent directement des œfs; tantôt ils donnent naissance à des mérides nouveaux qui peuvent, dès qu'ils sont formés, se séparer de leur parent et vivre d'une façon indépendante: c'est le cas de quelques éponges inférieurs, de l'hydre d'eau douce et d'un certain nombre de vers inférieurs...."

Or the mérides may remain conjoined in an organism of the third order of organization: "Les mérides nés les uns des autres peuvent aussi demeurer unis entre eux. Ils forment alors ce qu'on nomme de *communautés* ou des *colonies*." "The mérides of a colony may take on diverse forms and perform diverse functions; groups of mérides appropriate to these functions may detach from the colony ... and give rise to the remarkable phenomena of geneagenesis and alternation of generation. This is seen in the medusas and annelids. But also, all the clonal mérides may remain together, become modified in different ways, and become so interdependent ("solidaires") that they are inseparable; their ensemble then constitutes a new organism having all the characteristics of an individual. This is the case in all of the higher animals to which one can give the name of zooid [*zoïde*] or of *dème*....

But how does Perrier's concept of the *protoméride* relate to that of the méride? We see from p 196 of the present text that the former is the first-formed méride: "In vertebrates as in invertebrates, organic complexity has been obtained by more or less complete fusion of zoonites, that have budded one from another and the first - to which one may give the name *protoméride* – was alone originally the direct product of the egg."

Ontogeny of the protoméride proceeds according to whether it becomes sessile or remains mobile – when sessile, it becomes a radiate or zoophyte (Radiata or Zoophyta of Cuvier's system), when mobile, an arthropod or vertebrate. Given the crucial nature of this statement, we cite most of the original text (p 252): "Quand le protoméride se fixe, il produit par bourgeonnement de colonies irrégulaires, arborescents, ramifiées ou incrustantes ... pour produire des organismes rayonnés. Quand le protoméride demeure libre et rampant, il présente une symétrie bilaterale, ne produit de bourgeones qu'à son extrémité postérieure et donne naissance à des organismes segmentés dont les vers annelés, les arthropods et les vertébrés sont les principales formes. Les différents modes de symétrie qui caractérisent les grands types organiques trouvent donc une explication rationnelle, et il n'est plus nécessaire de fair intervenir directement une pensée créatrice distincte pour en rendre compte." So Perrier offers a rational explanation for the different modes of symmetry and the evolution of the major groups of organisms.

He has introduced the concept of what he calls protoméride and méride, as archetypes for what become 1) larvae and zooids (polyps) in the case of the radiate or zoophytic metazoa and 2) larvae and germs (embryos) and zoonites (= somites and metameres) in the articulates and vertebrates. The linguistic similarity of zooid and zoonite may not be accidental. He seems to have found this grand rational scheme quite satisfying as a mechanistic alternative to the idea of special creation ("pensée créatrice distincte'). However, his conceptualizations and terminology have given way, as with Geothe's "Urblatt," to less general terms for specific observable structures.

Metagenesis and alternation of generations (see also, Metamorphosis) Perrier states that Richard Owen used *metagenesis* for *asexual reproduction*, q.v., but the usual definition of this word is "The alternation of sexual and asexual generations in animals." Perrier credits Steenstrup with recognizing the phenomenon and coining this expression (see Perrier, Ch 17, p 175). In the Hydrozoa, for example, the course of ontogeny is from fertilized egg (zygote, with two sets of chromosomes), by mitotic cell division to a colony of polyps. Some of these (gastrozooids) function as feeding structures, others (gonozooids) are reproductive in that they bud off (asexually) medusae, which float away free from the sessile colony. The medusae are the site of reduction division (meiosis), that results in formation of the gametes (sperms and eggs). These latter close the life cycle by fusing and so creating the diploid zygote with its two sets of chromosomes. Thus in metagenesis, both sexual and asexual generations are diploid. This sort of alternation of generation stands in contrast to that of most plants. In their case, the cells that result from reduction division (meiosis) do not directly form gametes, but rather, proliferate mitotically and grow into a multicellular body (the haplophase). Gametes are produced by this generation by mitotic (equational) division of cells. Thus in most plants, the generations that alternate differ in their chromosome complements (haploid and diploid). Zoologists and botanists go their own ways with regard to terminology.

The following quotation (under *zooid* in the Oxford English Dictionary, Second Ed., 1989) explains clearly the dimorphism of polyps or zooids in the metagenetic life cycle of hydroids: "1871 T. R. JONES *Anim. Kingd.* (ed. 4) 98 In each colony [of hydroids] the alimentary and reproductive functions are respectively intrusted to two distinct kinds of zooids,..the nutritive and the sexual polypites... The sexual zooids, like the flower-buds of plants, are only developed at certain seasons."

Metamere, *n.*, A segment of that animal body or plant shoot, the structure of which is based on a linear array of segments (also called *somite* and *zoonite* in animals).

Metamorphosis (see also Metagenesis), *n.*, The current meaning of this word in biology is given in the Oxford English Dictionary, Draft Revision, June 2008: "Change of form in an animal (or plant), or its parts, during post-embryonic development; spec. the process of transformation from an immature form to a different adult form that many insects and other invertebrates, and some vertebrates (e.g. frogs), undergo in the course of maturing." By the year 1665, one referred to the metamorphosis of the silkworm insect, which meant its transformations in stages from larva to adult. By 1859 the distinction between metamorphosis (which takes place within one individual animal or plant body and metagenesis (which takes place between separated individuals, albeit genetically identical) was clearly made: "1859 Todd's Cycl. Anat. & Physiol. V. 13/1. This is the 'Alternating Generation' of Steenstrup, or what we might with

Mr. Owen call, in contrast to Metamorphosis, a process of Metagenesis." (Oxford English Dictionary, Draft Revision, June 2008).

Monad, *n*., (see also Plastid) This word has been used in three senses: 1) A single-celled organism. 2) A flagellate protozoan of the genus Monas. Or 3) "In the early quotations such organisms are often regarded as the lowest members of the hierarchy of nature or as fundamental units of the animal body." (Oxford English Dictionary).

Monism, *n*., "[T]he monism that was the basis of Haeckel's philosophy" [Perrier, Ch 14, p 10] is best understood from two quotations from the Oxford English Dictionary: "1876 E. R. Lankester tr. Haeckel Hist. Creation I. ii. 35 The Monism here maintained by us is often considered identical with Materialism." And "1984 E. Pawel Nightmare of Reason v. 69 Haeckel, who mounted Darwin's theories into the framework of a mechanistic monism that left no room even for Spinoza's monistic deity."

Mysis, *n*., 1) a crustacean genus of the order Mysidacea – opossum shrimps. 2) one of the stages of larval development within the decapod crustaceans. Simply put, this ontogenetic sequence is as follows: egg → nauplius → zoea → mysis → adult.

Nauplius, *n*., a crustacean larva, the first stage after leaving the egg, usually with three pairs of appendages, a median eye, and little or no segmentation. (See Mysis).

Need, *n*., The common English translation of "besoin" as used by Lamarck, is "need." Lamarck thought that an external physical challenge induces a response intrinsic to the organism that he called "need" and that this latter, together with habitual use of associated faculties, enables the organism to so transform itself so as to adapt to changing conditions. Some of Lamarck's opponents mistakenly used the word "wants" (desires) to imply that Lamarck intended there need be nothing more than a willful desire on the part of the organism to achieve its adaptation to new conditions. This distinction was clearly made by A. S. Packard (*Lamarck, The Founder of Evolution*. 1901. Longmans Green, and Co. p. 350, ff.). Packard, quoting from Cleland, states: "Lamarck gave great importance to the influence of new wants [i.e. needs] acting indirectly by stimulating growth and use. Darwin has given like importance to the effects of accidental variations acting indirectly by giving advantage in the struggle for existence." So in current terms, Lamark was emphasizing exercise or use of a faculty, that enhances its phenotypic expression. The attribute, being expressed, then could be acted on by selection (natural or artificial), and consequently, the genetic faculty for expressing the phenotypic trait could be transmitted. The individual lacking this genetic predisposition, might not leave as many descendants as that which had it.

Parthenogenesis, *n.*, development of a seed or embryo without fertilization by a male.

Pennatule, *n.*, a type of free-swimming polyp distinguished by a sail-like plume.

Phalanger, *n.*, a squirrel-like marsupial found only in Australia, New Guinea and adjacent islands. Most are nocturnal and arboreal. The koala is an atypical example.

Plastid, *n.*, Haeckel (and Perrier) used this term for cell. This usage is now obsolete (Oxford English Dictionary.) We translate it as "cell" here to avoid its being confused with "plastid" as it is used in botany to refer to organelles within the plant cell, cf. proplastid, chloroplastid etc. See also: Monad and Anatomical element.

Preformation, *n.,* Perrier's history of biological concepts is often a history of the shedding of ignorance that is conflated with religious belief. Assiduousness and refined techniques led people to discover ever more obscure or minute details of organisms. But still today the argument over divine intervention continues. The Oxford English Dictionary, in the following quotation, provides evidence of the linkage of the word preformation with religious belief: "1732 Hist. Litteraria IV. 195 It is easy to think that the Soul is a divine Automaton, still more wonderful, and that by a divine Præformation it produces these beautiful Ideas." This dictionary's second definition of the biological usage tells it all: "2. Biol. and Hist. Sci. The supposed existence of all the parts of an organism in rudimentary form in the egg or the seed (cf. ovism n., spermatism n.). Contrasted with epigenesis n. Freq. in theory of preformation. The doctrine was current in the 17th and 18th centuries, and was advocated by scientists such as Malpighi, Leeuwenhoek, Leibniz, and Réamur. It came to be associated especially with the naturalist Charles Bonnet (1720–93), following publication *of his* book Considérations sur les corps organisés (1762)."

This concept of preformation is well illustrated by the following quotation from Perrier, Ch 6, p 26: "Thus, Bonnet pronounced himself opposed to all attempts at mechanistic explanations of the animals; he showed himself to be a resolute adversary of epigenesis and maintained that all living creatures came from a germ cell that was already organized."

Perrier also refers to the doctrine of preformation as "The hypothesis of preexisting germ cells…" (Ch 6, p 37).

Principle of connections (Principe des connexions), *n.,* Geoffroy Saint-Hilaire elaborated this method for discovering equivalences (homologies) among structures in animals: one compares structures in similar relative positions and examines the relations and interdependencies among them. Two quotations from his "Philosophie Anatomique .... (1818)" exemplify this definition: "Or il est évident que la seule généralité à appliquer dans l'espèce est donnée par la position, les relations et les dépendances des parties, c'est-à-dire, par ce que j'embrasse et ce que je désigne sous le nom de connexions [p xxv];" and "... un organe est plutôt altéré, atrophié, anéanti, que transposé [pp xxxiii and 405]."

Protoméride See: Méride.

Race, *n.*, as used by Perrier, race corresponds to Darwin's *subspecies.*

Rayonné, *n.*, Perrier probably used this word as it is defined here: "Les rayonnés, animaux sans vertèbres dont les organes sont disposés en rayons autour d'un centre où d'un grand axe. (Émile Littré, Dictionnaire de la langue française (1872–1877)). I.e., The Radiata, invertebrate animals in which the organs are disposed radially around a center or major axis.

Rotifer, *n.*, any of the approximately 2000 species of microscopic invertebrates that constitute the class Rotifera, or Rotaria (phylum Aschelminthes).

Salpes, *n.*, a barrel-shaped, free-floating tunicate that moves by pumping water through its gelatinous body by means of contraction, and strains the water to feed on phytoplankton.

Sarcode, *n.*, the name given by Felix Dujardin (1835) to the membrane-bounded, amorphous living substance of protozoan cells. Later, the word was superceded by the name, "protoplasm" (from Hugo von Mohl, who studied cells of plants). Protoplasm, in turn, became differentiated into cytoplasm, nucleoplasm and substituent entities. See: Anatomical element.

Selachians, *n.*, a group of fish with laminated gills. It includes sharks and rays.

Siphonophore, *n.*, an order of the Hydrozoa, a colonial class of marine invertebrates belonging to the phylum Cnidaria. The colonies can have a superficial resemblance to jellyfish, and they are among the most abundant predators. The best known species is the Portuguese man o'war.

Somite See: Metamere.

Syllis, *n.*, a name coined by Savigny in 1818 for small (less than 2 inches) and very slender worms, on the order of 25-50 times longer than thick. Also called syllid.

Taredo, *n.*, ship-worm, a marine mollusc with an atrophied shell and vermiform body. It digs tunnels into the wood of ships. They reach 20 cm.

Tegument, *n.*, outer covering of the body, such as skin, hide, or the shell of crustaceans.

Teleology See: Cause.

Tenia, *n.*, tapeworm.

Teratology, *n.*, the study of anatomical anomalies or monstrosities.

Transformism (Transformisme), *n.*, By the 1870's, the word had taken on its modern, technical meaning of "evolution" (Émile Littré, Dictionnaire de la langue française, 1872–1877).

"Transformisme" is defined in the Larousse encyclopedia (2006) as the "Theory of the succession of fauna and flora in the course of geological time based on the idea of progressive transformation of the populations and offsprings either under the direct influence on individuals (Lamarck) or through modification or mutation following natural selection (Darwin, De Vries). In the latter sense, one may also say *evolutionism.*" In the present text it has been translated as *theory of organic evolution.* Darwin used the term *descent with modification.* Perrier, like most French naturalists of his time, rarely used the word evolution, because the term had various other meanings in French and did not always convey the same meaning it has in English. He also tended to use *natural selection* rather than *survival of the fittest.* Although Darwin did not use the latter until it was coined by Herbert Spencer (1820–1903), he commented that it was a more accurate expression of the principle he discussed.

Perrier also spoke of those who accepted the concept of transformism as transformists, for example, Buffon, Lamarck, Geoffroy Saint-Hilaire, and Goethe.

Trematode, *n.*, a group of un-segmented, parasitic flat-worms, including the flukes.

Trochophore, *n.*, a free-swimming ciliate larva typical of marine annelid worms but occurring in several invertebrate groups.

Tunicate, *n.*, a subphylum of marine chordate animals that have clefts in the vascular walls of the pharyngeal gills, a thick secreted covering layer, a greatly reduced nervous system, and a heart that can reverse the direction of blood flow.

Unity of plan of composition or constitution The phrase, "L'unité de plan de composition" was coined by Geoffroy Saint-Hilaire to describe the similarity of morphological and anatomical plan that he found among animals. This eventually led him to coin the phrase, "Theory of analogues." Geoffroy's "analogues" became Owen's "homologues" in the latter's "Hunterian Lectures" in 1843, in which, in its Glossary, he distinguished between "homology" and "analogy" in the context of biology. The kernel of the idea of homology is traced by Perrier from Aristotle to Linnaeus, Goethe and many others right to Owen. Truly, this idea is at the core of biological philosophy. But the idea of unity of plan is advanced as well in philosophical and theological contexts and inheres in many phrases in many languages.

First, comes the recognition of patterns of resemblances. Then follow explanations for them. Typology is involved; archetypes are conceived. Ultimately, the intellect struggles to explain the patterns either in terms of physical mechanisms or that of metaphysical force. Natural theology employs the patterns to reveal the hand of a Creator (cf. the argument of intelligent design). Biology explained them as the effect of causal forces: (natural) selection acting on a diversity of genetically modulated phenotypes.

Darwin used the phrase "unity of type" for "unity of plan:" "It is generally acknowledged that all organic beings have been formed on two great laws – Unity of Type, and the Conditions of Existence. By unity of type is meant that fundamental agreement in structure which we see in organic beings of the same class, and which is quite independent of their habits of life. On my theory, unity of type is explained by unity of descent [Darwin. On the Origin of Species. N.Y. D. Appleton & Co. 1883, p. 166]." Darwin appreciated fully Geoffroy's recognition of a "general plan" and "similarity of pattern" of organization: "Geoffroy St. Hilaire has strongly insisted on the high importance of relative position or connexion in homologous parts; they may differ to almost any extent in form and size, and yet remain connected together in the same invariable order." [Darwin, op cit. p. 382].

Darwin, as did St. Hilaire, favored a mechanistic, physical explanation for unity of type when he said, "It is so easy to hide our ignorance under such expressions as 'the plan of creation,' 'unity of design,' &c., and to think that we give an explanation when we only re-state a fact." [Darwin, op. cit. p. 422] Yet, Perrier, for all that he, too, advocates physico-chemical explanations for unity within diversity, shows that he understands the limits of knowledge – in his last words at his book's end: "... il convient de rappeler que la chimie, la physique, l'astronomie, en expliquant les faits qui appartiennent à leurs domaines respectifs, n'ont nullement atteint la cause première. La biologie moderne n'atteint pas davantage cette cause; elle ne supprime pas Dieu; elle le voit plus loin et surtout plus haut." "... it is helpful to recall that chemistry, physics, and astronomy, in explaining the factual evidence on which their respective domains are founded, have yet to discern a first cause. In this respect, modern biology is no different. It does not suppress God; it sees Him as more remote and, above all, higher."

Vitellus, *n.*, the yoke of an egg.

Zoea, *n.*, a larval stage in development of decapod crustaceans. (See Mysis).

Zooid, *n.* Reading the definition of this noun in the Oxford English Dictionary helps one to understand and interrelate some of the terms that pertain to life cycles and development, especially of the lower animals (e.g. Coelenterata): "Something that resembles an animal (but is not one in the strict or full sense): in early use applied somewhat widely, including, e.g., a free-moving animal or vegetable cell, as a spermatozoon or antherozooid; but chiefly restricted to an animal arising from another by asexual reproduction, i.e. budding (gemmation) or division (fission); spec. (and most usually) each of the distinct beings or 'persons' which make up a compound or 'colonial' animal organism, and often have different forms and functions, thus more or less corresponding to the various organs in the higher animals. (Cf. Zoon)" The word "zooid" lives on as a suffix in gastrozooid – the nutritive or alimentary polyp or zooid- and gonozooid – the sexual or gonad-bearing polyp. See also, méride and metagenesis.

Zoonite See: Metamere

Zoophyta, Zoophyte, *n.*, major classes in the classifications of animals by Linnaeus and Cuvier. Libbie Hyman in her classic study, succinctly explains their usages: "Linnaeus in 1758 divided the animal kingdom into the classes Mammalia, Aves, Amphibia, Pisces, Insectra and Vermes, thereby indiscriminately throwing all the invertebrates except insects into the group of Vermes or worms. This class he subdivided into the orders Intestina, Mollusca, Testacea, Lithophyta, and Zoophyta .... In Cuvier's system (1816) the invertebrates were divided into Mollusca ... Articulata ... and Radiata or Zoophyta." (Hyman, 1940. pp 27–28). When, as on one occasion, Perrier places Zoophyta within quotation marks, he is referring to them as a set of animals with indeterminate and open growth development, similar to that of plants.

Biographical Index

The following biographical notes are intended to provide a convenient source of basic information on the principal individuals referred to in the text.

Agassiz, Louis (1807–1873). Swiss paleontologist and glaciologist. After attending the universities of Zürich, Heidelberg, Munich and Erlangen he went to Paris to study under Georges Cuvier, who encouraged the studies of ichthyology that became one of Agassiz's life-long pursuits. After Cuvier's death, Agassiz obtained a professorship at Neuchatel in Switzerland, where he developed the interest in glaciology that eventually gained him the title "Father of Glaciology." While in Switzerland he had opportunities to observe the effects of modern glaciers and after seeing similar glacial features throughout Germany and Britain concluded that a vast sheet of ice had recently extended over much of northern Europe and the British Isles. The publication of his observations in 1837 discredited the prevailing theories that attributed the wide-spread glacial deposits to a great flood.

Agassiz came to the United States in 1846 to explore the glacial geology of North America and was appointed professor of zoology and geology at Harvard University the following year. In 1859 he founded the Museum of Comparative Zoology and in 1863 was a founding member of the National Academy of Sciences.

In his *Essay on Classification*, first published in 1851, Agassiz followed Cuvier's general classification of life and noted that fossil remains are more primitive in the oldest rocks and become more developed in younger strata. He used this observation to place *homo sapiens* at the summit of the scale of life. In a later (1869) edition he attacked Darwin and maintained that life followed a Divine Plan. He was disappointed that most of his students – including his son Alexander, a well-known naturalist in his own right – accepted some form of evolution.

Agassiz argued that species were distributed where God meant them to dwell. His views on ethnology and anthropology were explicitly opposed to Darwin's argument that the human races had a common origin and that racial differences were only superficial. Darwin's second book on evolution, *The Descent of Man*, explicitly addressed Agassiz's theories. In later years, Agassiz was widely criticized for his racist views on the superiority of the white race. He died in Cambridge,

Massachusetts on the 14th of December 1873. Edward Lurie (1960) has given us an excellent account of his life and works: *Louis Agassiz: A Life in Science,* University of Chicago Press.

Aldrovandi, Ulisse (1522–1605), Italian renaissance naturalist and professor at the University of Bologna, notable for his systematic, precise observations of animals, plants, and minerals. He was the first to describe the progressive changes during incubation of chicken embryos. His collection of biological specimens, which he classified according to his own system, contributed to the later development of animal taxonomy.

Allmers, Hermann (1821–1902), a German writer and poet. In his *Marschendichter* (1858) he described the culture and geography of his native northwestern Germany.

Anaximander (611–547 BC), a Greek philosopher, was probably the first to express the concept of evolution. He thought that humans evolved from creatures in the sea.

Anaxagoras (500–428 BC), a prominent philosopher in Athens during the age of Pericles, attempted to explain natural phenomena, such as eclipses, rainbows, and the solar system in terms of physical rather than supernatural principles. His interest in meteorites led him to propose new theories for the origin of the universe. He proposed that the constituent elements of life have existed throughout all eternity and that there seemed to be no essential differences to distinguish living from inanimate material.

Aristotle (384–322 BC), was probably the single most influential philosopher of the Western World. He studied with Plato at the Academy in Athens for 20 years (367–347 BC) before becoming tutor to Alexander of Macedonia (347–335 BC). On returning to Athens, he became head of the Peripatetic School in the Lyceum (335–322 BC). For centuries, his work dominated thinking on physics and biology, as well as logic, ethics, poetry and drama. His rules of causality, described under *Final Causes* in the Glossary, were the subject of much debate, especially during the Middle Ages and Renaissance.

In his *History of Animals,* Aristotle placed all beings in a *scala naturae* ranging from plants to humans. Plants can reproduce and grow, animals reproduce and grow but are also mobile and sensitive, while humans also have the capacity of thought. He recognized similarities in the anatomic structure of animals and pointed out that the fin of a fish, the wing of a bird, and the arm of a man were analogous. Starting in the 11th century, Albertus Magnus, Duns Scoticus, and Thomas Aquinas "Christianized" Aristotelian logic in an attempt to reconcile reason with faith, but when modern science began to take form in the 16th century it brought a revolt against Aristotle's central role in mediaeval dogma. It was not until the 19th century that the importance of his biological insight gained the recognition it deserved.

Bacon, Francis (1561–1626) was an English philosopher, statesman, and very prolific essayist. While at Cambridge University he became dissatisfied with the teaching that still followed the work of Aristotle and went on to propose a *Scientific Method* based on rigorous inductive reasoning from factual evidence. The principles laid out in his best-known work *Novum Organum* (1620) made him a champion of modern science.

Bacon, Roger (1214–1292), an English philosopher and Franciscan friar, advocated empirical science and experimentation and wrote on a broad range of subjects including mathematics, physics, and astronomy. He attempted to introduce science into the teaching of the Franciscan Order.

Baer, Karl Ernst von (1792–1876), a German geneticist and embryologist, discovered that the eggs of mammals are produced in the ovaries. He rejected preformation but supported Cuvier's views on the immutability of species.

Belon, Pierre (1518–1564) wrote a *Histoire naturelle des animaux marins* and a *Histoire des oiseaux*. When he compared the organs of various animals he opened the way for comparative anatomy. Early in his studies of ornithology he presented a drawing of a bird's skeleton beside that of a human and designated with the same letters the parts that seemed to him to correspond in the two skeletons.

Bichat, Xavier (1771–1802), a French anatomist and physiologist, contributed to general anatomy and the development of histology. He conducted experiments designed to explore the workings of the senses.

Blainville, Henri de (1777–1850), a French zoologist, inherited Cuvier's chair for comparative anatomy at the Muséum d'Histoire Naturelle. Although a student of Cuvier, he opposed many of his master's ideas. He was also openly hostile to the work of Geoffroy Saint-Hilaire. Best known for his work on metabolism and ordering of animal life, he believed all animals had been created simultaneously but had developed gaps when certain species became extinct.

Blanchard, Emile (1819–1900), a French entomologist, occupied the chair for Crustaceans, Arachnids, and Insects at the Muséum d'Histoire Naturelle from1862 until 1895. He is best known for his ordering of the animal kingdom.

Bojanus, Ludwig Heinrich (1776–1827), a German physician, naturalist, and professor of comparative anatomy at the University of Vilnius, is best know for his work on turtles.

Bonnet, Charles (1720–1793), a Swiss naturalist with broad interests, is best known for his innovative experiments and innovative thoughts on evolution. He discovered parthenogenesis (development of the female's egg without sperm) in certain insects, described the regeneration of lost parts in freshwater hydra, and showed that the respiration of caterpillars and butterflies was through pores in their bodies rather than lungs. He devised experiments that demonstrated that the leaves of plants released oxygen through the process of photosynthesis. He was the first to use the term evolution in a biological context and had an important influence on Erasmus Darwin and his grandson Charles. Bonnet followed Aristotle and arranged the forms of life in a "Ladder of Life" ranging from the most simple to the most complex, namely humans.

Brongniart, Alexandre (1770–1847), French geologist and paleontologist. Together with Cuvier, he compiled a geological map of the Paris Basin that showed an ordered succession of Cenozoic fossils. This is considered the first geological map based on a stratigraphy defined by fossil remains.

Buffon, George-Louis Leclerc, Comte de (1707–1788), a very prolific French naturalist with exceptionally broad interests. While charged with management of the Royal Gardens, now known as the Jardin des Plantes, he organized scientific research in what later became the museum of natural history.

Although Buffon began his career as a strict creationist, he became convinced of the importance of evolution when he noted the presence of vestigial organs in creatures that must have evolved from earlier forms under the influence of changing environmental conditions. Although he rejected the Biblical concepts of creationism, he believed that there was a grand plan that governed all forms of nature and that the different forms of life were derived from a fixed number of ancestral forms that originated by spontaneous generation in the early history of the Earth and diverged by adaptation to changing environmental conditions. Depending on the nature of these conditions they could either progress to higher forms or degenerate. He recognized that many diverse types of animals share a similar anatomical structure.

His written work contained many of basic concepts of evolution and natural selection that were picked up by his followers at the Museum of Natural History, especially Lamarck and Geoffroy Saint-Hilaire. He also had a strong influence on Charles Darwin who remarked in the introduction of the *Origin of Species* that Buffon recognized the concept of evolution and was 'the first author who in modern times has treated it in a scientific spirit.'

Carus, Karl Gustav (1789–1868), a German painter, naturalist, psychologist and medical doctor. He was professor of obstretrics and director of the school of medicine and surgery at the University of Dresden.

Chambers, Robert (1802–1871), a Scottish publisher and member of the Royal Society of Edinburgh, was the author of the popular book *Vestiges of the Natural History of Creation,* which he published anonymously in 1844. Although severely criticized as immoral by theologians and as amateurish by scientists, it was very influential in bringing evolution to the attention of the general public and paved the way for Darwin's *Origin of Species* fifteen years later. Darwin saw it as a warning of how much opposition his evolutionary theories were likely to encounter.

Cuvier, Baron Georges Leopold Frédéric (1769–1832) was a dominant figure in the natural sciences during the late 18th and early 19th centuries. Born in a French-speaking Lutheran enclave that was then part of the Duchy of Württemberg, he studied zoology in Stuttgart under Friedrich Kielmeyer. He later took a job as tutor for the d'Héricy family in Normandy, where he was able to pursue his scientific interests and soon attracted the attention of natural scientists in Paris. In 1802 he obtained a chair in comparative anatomy at the Museum of Natural History. In 1812, he proposed a classification of animal life consisting of the four major divisions or *embranchements*: vertebrates, mollusks, articulates (insects, crustacea, etc.), and radiata (starfish, jellyfish, etc.). His explanation for the great variety of species found in nature was based on their "conditions of existence," which were governed by a higher power. The Creator designed species to live under these conditions, and if the conditions changed, the species could become extinct. He recognized the progression of species with time but believed that they become extinct as a result of global catastrophes and are replaced by newly created species that are better suited to the new conditions. Although he avoided religious interpretations of nature and preferred to stress systematic factual observations, his creationist views brought him into conflict with his close friend and colleague at the Museum, Etienne Geoffroy Saint-Hilaire. While Cuvier considered his primary responsibility to be collecting and classifying data and avoided what he considered unwarranted speculation, Geoffroy used factual evidence as a foundation upon which to develop theoretical insights. In 1820, their differences resulted in a widely publicized controversy that was essentially a debate over evolution versus creationism.

Cuvier cited Alpine deformation as evidence that violent earth movements were responsible for catastrophic events that could also account for the sudden extinction of species on a global scale. He reasoned that the well-preserved bodies of mammoths found in the Siberian tundra testified to a former tropical regime in that region and that the climate must have suddenly changed. Perrier, following the views of Lyell, Hutton, and others, laid out the arguments against this interpretation: all geological features could be explained in terms of the slow but steady effects of observable processes. It was not until the end of the twentieth century that the impact of comets and meteorites was seen as the probable cause for many mass extinctions and Cuvier's concept regained respectability in a new form.

Martin Rudwick's *Georges Cuvier, Fossil Bones, and Geological Catastrophes* (1997) gives a succinct account of Cuvier's career along with translations of some of his most important publications. Toby Appel gives a very readable account of this in *The Cuvier – Geoffroy Debate: French Biology in the Decades Before Darwin* (1987).

Cuvier, Fréderic (1773–1838), French zoologist, younger brother of Georges, author of a *Histoire des cétacés,* and collaborator with Geoffroy Saint-Hilaire on *Histoire des mammifères.* He occupied the chair of Comparative Physiology at the Muséum d'Histoire Naturelle from 1837 until his death the following year.

Darwin, Charles (1809–1882) developed his interests in natural history while studying medicine at the University of Edinburgh and theology at Cambridge. He was especially attracted to geology by the work of Charles Lyell. During the five years he served as geologist on the round-the-world voyage of the HMS Beagle, many of his perceptive observations on the volcanic islands of the Atlantic Ocean and the Galapagos lent support to Lyell's principle of *uniformitarianism*. Contrary to the popular view, he did not appreciate the variations among the Galapagos finches until two years after his return to Britain when John Gould helped him sort out his poorly documented specimens. (See Sulloway, 1982a & b) He delayed publishing his work for many years, and it was not until Charles Lyell alerted him to Alfred Russel Wallace's pending publication of a book advocating the principle of natural selection, that he agreed to share priority with Wallace. Finally, in 1859, twenty-four years after the return of the Beagle, he published his landmark *Origin of Species*. Although many earlier naturalists had postulated similar concepts of evolution, his lucid explanations and convincing examples won over most, but by no means all, of the scientific community. He met with fierce opposition from creationist, such as William Buckland and Richard Owen, who continued to adhere to a strict Biblical interpretation.

Darwin, Erasmus (1731–1802) was an English anatomist, pathologist, poet, and grandfather of Charles Darwin. His major scientific work, the two-volume *Zoonomia* (1794–1796), was devoted primarily to pathology but also included a number of insightful observations on biology and evolution. He does not seem to have recognized the importance of natural selection. Some of his hypotheses had an uncanny resemblance to those proposed at the same time by Lamarck, and it has been suggested that either he or Lamarck was guilty of plagiary. As Osborn (1922, p 152–5) shows, the accusation is probably unfounded.

Daubenton, Louis (1716–1800) was a French naturalist with unusually broad interests ranging from biology to mineralogy. Buffon enlisted his help in compiling the anatomical descriptions of quadrupeds in a multi-volume treatise on natural history. This led to a series of later publications on the comparative anatomy of animals, both living and extinct, as well as the physiology of plants. He also carried out experiments for the improvement of agricultural stocks and is credited with the introduction of merino sheep, which proved highly successful in France. He held a post as professor of mineralogy at the Jardin des Plantes and was a popular lecturer on many diverse topics. His work is said to have had an important influence on Geoffroy Saint-Hilaire and contributed to the latter's unified plan of animal anatomy.

De Maillet, Benoit (1656–1738), a French naturalist, sought explanations for the morphological development of the Earth over what he concluded must have been a much longer span of time than the biblical account. He offered an early hypothesis for the origin of man and proposed that cells are the basic unit of all forms of life.

His writings under the name Telliamed (De Maillet spelled backwards) were very influential.

Democritus (460–370 BC) suggested that individual organs could adapt to prevailing conditions. Cuvier called him the first comparative anatomist.

Diderot, Denis (1713–1784), a prominent figure in the French Enlightenment, is best known as the editor of the *Encyclopédie* and author of many wide-ranging philosophical works. His essay, *Lettre sur les aveugles*, described the dependence of animal life on the five senses and proposed a method by which the blind could be taught to read through their sense of touch. An early advocate of the variability of animal species, he recognized the importance of natural selection. His *Encyclopédie, ou dictionnaire raisonné des sciences, des arts et des métiers* includes articles by many notable scientists and philosophers on the new ideas and inventions that were emerging at that time. The first volume appeared in 1751 and the last in 1772.

Dugès, Antoine (1797–1838), professor of pathology and operative medicine at the Faculté de Médicine in Montpellier, attempted to reconcile the conflicting views of Cuvier and Geoffroy Saint-Hilaire. While generally convinced by Geoffroy's work, he sought to take Cuvier's objections into account by making Geoffroy's criteria for the "unity of organic composition" more flexible. In his *Memoire sur la conformité organique dans l'echelle animale* (1832), he proposed that animals are composed of multiple organisms that, in the early stages of development, were capable of living independently but gradually coalesced and became modified in order to conform to the functions of the body as a whole.

Dujardin, Felix (1801–1860), a French naturalist, described cellular cytoplasm and proposed that all single-celled animals should be considered as a separate group which he call Rhizopoda – now known as Protozoans.

Dumas, Jean-Baptiste (1899–1884), a French chemist, determined the atomic mass of a great many elements. His studies of the biochemical character of plant and animal life provided important insight into the processes by which life has evolved.

Empedocles (495–435 BC) was one of the first to express the idea of natural selection. Aristotle's rejection of his theory may have prevented him from receiving the recognition he deserved.

Fabre d'Olivet, Antoine (1767–1825), a French poet, playwright, and Biblical scholar, attempted to reinterpret the book of Genesis but his work was discredited and condemned by the Pope.

Flourens, Pierre, (1794–1867), Professor of Human Anatomy at the Muséum d'Histoire Naturelle, was a disciple of both Cuvier and Geoffroy Saint-Hilaire and attempted to combine their individual contributions in a compatible manner. He carried out experiments to determine the role of the different sections of the brain in governing behavior and the response to sensations.

Galen, Aelius Galenus (AD 129–200), a Roman physician and philosopher whose medical theories prevailed for more than a thousand years, especially in the Islamic world. He was a strong believer in experimentation and empiricism.

Geoffroy Saint-Hilaire, Etienne, (1772–1844) studied theology and law at the College of Navarre in Paris but his passion for science was aroused by the lectures on mineralogy by Daubenton and botany by Jussieu. Instead of taking up the priesthood, he switched first to medicine and then to mineralogy, which he studied under Abbé René-Just Haüy (1743–1794), the founder of modern crystallography.

In 1792, following the uprising of August 10th, Haüy, along with many other churchmen, was incarcerated in the Saint Firmin seminary, which was being used as a temporary prison. With the help of Daubenton and other scientists, Geoffroy succeeded in obtaining Haüy's release. Fearing for the fate of other churchmen still imprisoned in the building, Geoffroy disguised himself as a prison guard and used a fake identity card to gain entrance to the building where he tried unsuccessfully to lead the captives out through a back passageway. That night, after seeing one of the priests thrown to his death from a window, he obtained a long ladder, climbed to a second-story window, and led twelve elderly priests to freedom.

Greatly impressed by the young man's courage and initiative, Haüy enabled Geoffroy to switch from medicine to natural science and obtained a post for him at the Jardin des Plantes. When the Muséum d'Histoire Naturelle was created, Geoffroy became the first professor of vertebrate zoology and. together with Lamarck and Cuvier, went on to make the Museum the leading center for natural sciences.

Geoffroy was a member of the team of scientists who accompanied Napoleon on his Egyptian campaign, and the unusual forms of life he observed there awakened his interest in zoology. He was captivated by the idea that animals shared a uniform anatomical structure, and as he pursued his studies, the concept became the central theme of his work. This brought him into conflict with his close friend and colleague, Georges Cuvier, who fiercely resisted any possibility that living forms could change or adapt to ecological conditions. The ensuing debate is described in Appel's *The Cuvier – Geoffroy Debate: French Biology in the Decades Before Darwin* (1987). Geoffroy was developing a system in which all animal life could be related to a single primitive form when, in 1840, a stroke left him blind and paralyzed. The following year he passed his post at the museum on to his son Isadore and died three years later. Hervé Le Guyader's excellent account of his life and work has been translated from the French by Marjorie Grene (*Geoffroy Saint-Hilaire, A Visionary Naturalist*).

Geoffroy Saint-Hilaire, Isidore, (1805–1861), son of Etienne Geoffroy Saint-Hilaire. After receiving his degree in medicine, he became interested in malformations of the fetus, and his *Histoire générale et particulière des anomalies de l'organisation chez l'homme et les animaux* (1832–1837) made teratology a respected branch of science. After serving three years as professor of zoology at the University of Bordeaux, he returned to Paris in 1841 to take his father's post at the Museum of Natural History where he worked on his father's collections and clarified some of the latter's hasty observations. In his *Essaies de zoologie générale ou, Mémoire et notices sur la zoologie générale, l'anthropologie et l'histoire de la science* (1841), he made his father's work more acceptable by moderating some of his more extreme views.

Godron, Dominique Alexandre (1807–1880), a French botanist, zoologist and ethnologist, was head of the Faculty of Sciences at Nancy. He is best known for outlining the relationships of species and races, which he defined in his book "*De l'Espèce et des Races dans les Etres Organizés et Spécialment de L'Unité de l'Espèce Humaine.*

Goethe, Johann Wolfgang (1749–1832), known for his broad intellectual achievements in both literature and science, explored a wide spectrum of nature and human behavior. George Eliot called him "the last true polymath to walk the face of the earth." While attempting to revive an abandoned silver mine in the Harz Mountains, he developed an interest in geology and mineralogy and recognized the iron-oxide mineral goethite, one of the weathering products of sulfide minerals in ore deposits. He developed a theory of color and light that clashed with that of Newton but in later years was recognized as conceptually in advance of his times. He explained the morphology of plants and animals as the result of gradual metamorphosis according to a simple structural plan. In plants, the basic form was the leaf, and in animals it was the vertebra, both of which underwent transformations to form the principal parts of the body. In this respect he anticipated the work of Richard Owen, Geoffroy Saint-Hilaire, and Dugès and had an important influence on Darwin. Richard Friedenthal gives an excellent account of his work in *Goethe: His Life and Work* (1963).

Haeckel, Ernst (1834–1919), an early German supporter of evolutionary concepts, attempted to synthesize the ideas of both Lamarck and Darwin in a single philosophical system that would be compatible with Christianity. His book, *Natural History of Creation*, went through twelve editions and was translated into all the major European languages. His *biogenetic law* that "ontogeny recapitulates phylogeny" held that embryonic stages correspond to ancestral adults and provided direct evidence of lines of descent. He proposed that the nucleus of the cell transmitted hereditary traits. It is now known that the resemblance to progenitors is not found in adults but is restricted to the embryonic forms and that embryos evolve just as the adults do. Despite this modification, the law still has considerable heuristic value. He was accused, probably unjustly, of manipulating illustrations to show resemblances of the embryos of diverse forms of life. Haeckel is credited with introducing the word *ecology*. Robert Richards has given an excellent account of his life and work in *The Tragic Sense of Life: Ernst Haeckel and the Struggle over Evolutionary Thought*, 2008.

Heracleitus (c. 540–480 BC), a Greek philosopher, is remembered for his ideas on cosmology in which fire and thermal energy are the basic elements of the universe. He introduced the concept that many things are defined in terms of opposites: good and evil, health and disease, hot and cold.

Kielmayer, Karl Friedrich (1765–1844), a German anatomist and early member of the German school of *Naturphilosophie*, was one of Cuvier's teachers at Stuttgart and had a strong influence on the latter's thinking.

Kœlreuter, Joseph Gottlieb (1733–1806), a German botanist, carried out experiments on hybridization of plants and showed that the pollen contributed as much to the characteristics of the hybrid offspring as the parental seed.

Lacaze-Duthiers, Henri (1821–1901), a French zoologist, is best known for his work on mollusks.

Lamarck, Jean-Baptiste (1744–1829) was born in Bazentin, Picardy, where his family belonged to the minor nobility. After attending a Jesuit College in Amien, he joined the army and served with distinction in the Seven Years War. When poor health forced him to leave the service in 1768, he found a job in a bank in Paris and in his spare time attended lectures on medicine and a wide range of scientific subjects, including botany under Xavier Bichat. His ability and enthusiasm attracted the attention of George Buffon, who hired him to serve as a tutor and companion to his son during a tour of Europe. This led to a position at the Jardin du Roi, of which Buffon was then director. When the Museum of Natural History was established in 1793, Lamarck was given the chair for entomology, worms, and microscopic animals.

Lamarck became interested in the adaptation of plants to changing climatic conditions and recognized that the same general principles applied to all forms of life. (In 1802 he coined the word *biology* to include all forms of natural life.) Lamarck's interpretations of evolution (see discussion in the Glossary) clashed with the prevailing biblical doctrine and were not well received until the French revolution upset the authority of the church. Cuvier was particularly alienated, and although they were colleagues at the Museum, they were scarcely on speaking terms. When Cuvier noted that the mummified animal species that Geoffroy Saint-Hilaire brought back from Egypt were identical to modern species, he took this as evidence that species were immutable. Lamarck responded that it was wrong to expect any change in the species, because the ecological conditions of Egypt had not changed significantly since the time the mummified animal had lived there.

Some of his hypotheses had an uncanny resemblance to those proposed at the same time by Erasmus Darwin, and it has been suggested that either he or Lamarck was guilty of plagiary. As Osborn (1922, p 152–5) shows, the accusation is probably unfounded.

In his later years, Lamarck lost his eyesight and was cared for by his two daughters until he died in virtual poverty in 1829, a year before the Cuvier – Geoffroy debate brought his work to the attention of the general public. Lamarck had wide-ranging interests and tended to indulge in very speculative ideas when dealing with topics about which he had little direct experience. For example, he subscribed to the Wernerian idea that all crustal rocks, including granite, were precipitated from seawater, and beliefs of this kind tended to discredit his work as a whole. His only comprehensive published work, *Philosophie zoologique* (1809), did not give a very satisfactory account of his interpretations of evolutionary processes contributions, but the biographies by Jordanova (*Lamarck*, 1984, Oxford University Press) and Packard (*Lamarck, The Founder of Evolution – His Life and Work*, 1901, Longmans - Green) provide good summaries of his theoretical contributions.

Lesueur, Charles Alexandre (1778–1846) described thousands of zoological species from Australia, Tasmania, and North America. A talented artist, his illustrations are greatly admired.

Leuckart, Karl George Friedrich Rudolf (1822–1898), a German zoologist and parasitologist, studied the invertebrate marine life of the North Sea as well as the tapeworm and trichinosis.

Leuwenhœk, Anton van (1632–1723), a Dutch naturalist, introduced the use of magnifying lenses for studies of nature and revealed the existence of infusoria and spermatozoa. His recognition of the asexual reproduction of the plant louse, later verified by Bonnet, and his observations on the generation of polyps by budding were generally overlooked until Trembley drew attention to them.

Linnaeus, Carolus (1707–1778), Swedish taxonomist, whose proper name was Karl von Linné, systematized the naming and classification of plants, animals, and minerals according to a binomial system of Latin nomenclature in which the first (generic) name is shared by all members of the same genus and the second is specific to a species. In addition to genera, he established families, orders, and classes, but these were less successful, because they proved to have many exceptions. Nevertheless, his system facilitated the development of evolutionary thinking by clarifying morphological relationships. (See discussion in the Glossary.)

Linnaeus assumed that each species descended from a single, immutable pair that had been individually created by God according to a uniform, harmonious plan that made them ideally suited for the conditions in which they were placed.

Lucretius (ca. 99–55 BC), Roman poet and philosopher, left only one known work *De rerum natura (On the Nature of Things*). He sought to replace superstition, including religion, with a rational understanding of the universe.

Malpighi, Marcello (1628–1694), a professor of medicine at Bologna, is considered the founder of microscopic anatomy. In addition to drawing attention to a great number of peculiarities in the structure of human organisms, he discovered the windpipes of insects, and studied the development of the chicken.

Maupertuis, Pierre Louis Moreau de (1698–1759), a French mathematician and naturalist, spent much of his active life in the court of Frederick II of Prussia. He is credited with the Principle of Least Action, which was an early application of the laws of thermodynamics to physics and biology. He was a foremost advocate of the oblate form of the Earth, which he explained according to Newtonian principles and demonstrated by measuring a degree of longitude in Lapland and comparing it to another closer to the equator. He clearly enunciated the concept of survival of the fittest in his 1745 work *Venus Physique*, but did not elaborate it or offer examples.

Milne-Edwards, Henri (1800–1885), the 27th child of an English father and French mother, was one of the founders of modern French physiology. He was one of Perrier's colleagues at the Museum of Natural History where he succeeded Isidore Geoffroy Saint-Hilaire at the chair of Zoology. His work on mollusks and crustaceans set up the faunal groupings of marine life.

Moquin-Tandon. Horace Bénédict Alfred (1804–1863), French naturalist and medical doctor, studied invertebrate animals, particularly leeches. He held the post of Professor of Botany at the University of Marseille and was later director of the Jardin des Plantes in Paris.

Müller, Fritz Max (1823–1900), a German philologist, spent most of his active life at Oxford where he studied mythological accounts of natural phenomena. His book *Für Darwin* developed Haeckel's idea that embryology recapitulates phylogeny.

Naudin, Charles Victor (1815–1899), a French botanist, worked on hybrid plants and carried out studies similar to those of Mendel.

Oken, Lorenz (1779–1851), a German naturalist shortened his original name, Ockenfuss, to Oken. After studying history and medicine at the university of Würzburg he obtained a post at the University of Göttingen and soon became a leading figure in the school of *naturphilosophie*. In his seminal *Grundriss der Naturphilosophie* (1802) he laid out his basic thesis that "the animal classes are virtually nothing but a representation of the sense organs, and that they must be arranged in accordance with them." According to this principle he recognized five classes of animals: (1) invertebrates, that had only a sense of feeling through their skin, (2) fish, the animals in which a true tongue first appeared, (3) reptiles, which were the first to have a sense of smell, (4) birds, which were the first to have an external ear, and (5) mammals, in which all the senses were fully developed. In 1805, he proposed that all organic beings originate from and consist of *vesicles* or *cells* that are the protoplasm (*urschleim*) from which all larger organisms have evolved. His work came to the attention of Goethe, who offered him a special chair in medical sciences at the University of Jena. Shortly after accepting the post he published his interpretation of the bones of the skull as an extension of the vertebral column. This idea had already been proposed by Autenreith, Kielmeyer, and Frank.

Owen, Richard (1804–1892), an English naturalist and comparative anatomist, is best known for his system based on Kant's concept of "archetypes" that serve as unifying structural plans for the diverse forms of different groups of organism. He clarified the terminology of serial (within individuals) and special (between individuals) analogies. He was prominent in early studies of dinosaurs and is credited with founding of the British Natural History Museum, but he is said to have been guilty of unethical professional behavior and was a bitter opponent of Darwin and his theory of evolution. The biography by Rupke, *Richard Owen, Victorian Naturalist* (1994), gives a good account of his life and work, and Cadbury has described the more controversial aspects of his work in her book *Terrible Lizard* (2000).

Paley, William (1743–1805), an English churchman, sought to show in his *Natural Theology or Evidences of the Existence and Attributes of the Deity collected from the Appearance of Nature* (1802) that scientific studies could be used to prove the existence of God. The rigorous logic and broad scope of his scientific observations were intended to provide a bridge between the scientific view of the natural world and his theological beliefs. He is best known for his "Watchmaker Analogy." In his book *Before Darwin: Reconciling God and Nature* (2005) Keith Thompson gives a full account of the influence of his work on pre-Darwinian philosophy.

Pinel, Philippe (1745–1826), a French physician and psychologist, specialized in the treatment of mental illnesses.

Pliny, Gaius (AD 23–79), better known as Pliny the Elder, was a natural philosopher and author of *Naturalis Historia* in which he discussed a wide range of biological and geological topics. He died from asphyxiation while attempting to rescue friends during the 79 AD eruption of Vesuvius.

Prévost, Constant (1787–1856), a French stratigrapher, used fossil fauna to determine relative ages of geological formations and to define the paleogeography of Europe during the Tertiary Period.

Pythagoras (c. 580–500 BC), a Greek philosopher, is best known for his development of mathematics, including the Pythagorean Theorem. He believed that the stability of the universe is based on a balance of opposing forces.

Quatrefages, Jean Louis Armand de (1810–1892), a French naturalist, gave up his medical practice to pursue research and with the help of Henri Milne-Edwards, obtained a teaching position in Paris. In 1855 he was given the chair of anthropology and ethnology at the Muséum d'Histoire Naturelle. He was admired for his thorough studies of the anatomy of all levels of animal life and for the vigorous style of his writing on a broad range of topics. He was strongly opposed to Darwinian evolution.

Rathke, Martin Heinrich (1793–1860), a German anatomist and professor at the University of St. Petersburg, was one of the founders of modern embryology.

Ray, John (1628–1705) has been called the father of English natural history. The son of a country blacksmith, he grew up in a small town in Essex and at the age of sixteen went to Cambridge to prepare for the church. He developed a strong interest in natural science, languages and mathematics, and in 1651 became a lecturer at Trinity College. An unyielding puritan, he refused to accept the Act of Uniformity passed by the reactionary Cavalier Parliament in 1662 and was forced to give up his fellowship at Cambridge.

In his influential book *The Wisdom of God in the Creation,* Ray argued that God had not created species individually but had designed the world according to a system of natural laws. He did not believe that God would have allowed species to become extinct as the result of natural catastrophes. Instead, he maintained that survivors of life forms that seemed to have disappeared from the geological record would eventually be found in remote, unexplored parts of the world.

Rédi, Francesco (1626–1698), an Italian naturalist, used well-designed experiments to discredit the hypothesis of spontaneous generation. He conceded that this form of generation might be possible in the case of worms found inside fruit and in the intestines of humans and animals, but he argued that it was the vital forces themselves, the embryonic *souls of vegetation,* that engendered these worms.

Sars, Michael (1805–1869), a Norwegian anatomist, is best known for his studies of brain cells.

Schelling, Friedrich Wilhelm Joseph von (1775–1864), one of the leading philosophers of the German school of Naturphilosophie, was a protégé of Goethe. He held posts at the Universities of Wurzburg and Munich.

Schleiden, Mathias (1804–1881), Professor of Botany at the University of Jena, proposed a theory of spontaneous generation of primordial germs that developed into species. Together with Schwann, he is credited with the development of cell theory.

Schultze, Max Johann Sigmund (1825–1874), a German anatomist, was a professor at Halle University and later at Bonn where he was also the director of the Anatomical Institute. He developed an early theory of cell structure.

Schwann, Theodor (1810–1882), a German biologist and co-discoverer with Mathais Schleiden of the cell theory, showed the differences between cells of plants and animals.

Serres, Etienne Reynaud Augustin (1786–1868), was a French physician at the Hôpital de la Pitié and professor of human anatomy at the Jardin des Plantes where he was a disciple of Geoffroy Saint-Hilaire. Together with the German anatomist, Johann Friedrich Meckel (1781–1833), he proposed the "Meckel-Serres Law" that attempted to use comparative embryology to show that a single "unifying pattern" prevailed throughout the organic world. This was an early version of Ernst Haeckel's biogenetic law that "ontogeny recapitulates phylogeny".

In the field of teratology, Serres explained malformed organs as cases of arrested development or over-development. He had disagreements with Charles Darwin regarding the latter's evolutionary theories. Serres believed that humans were set apart from other forms of life as the supreme goal of all creation.

Steenstrup, Johannes Japetus (1813–1897), Danish biologist, professor of Zoology at the University of Copenhagen, described alternating sexual and asexual reproduction in parasitic worms.

Swammerdamm, Jan (1637–1680), a Dutch naturalist, introduced the use of magnifying lenses into studies of nature. His most important work, *Biblia naturæ,* is famous for its account of his research on the metamorphosis of insects.

Telliamed see De Maillet

Thales of Miletas (624–546 BC) was one of the first philosophers in the Greek tradition. Bertrand Russell considered him the "Father of Science." He made Miletas, a port on the coast of what is now Turkey, an early center for natural philosophy where Amaximenes and Pythagoras were among his many notable students. He was one of the first Greek thinkers to advocate a scientific approach to studies of life in the universe and tried to explain nature in terms of physical forces.

Trembley, Abraham (1710–1784), a Swiss naturalist, considered by some the Father of Biology, was among the first to utilize experimental methods to open a new view of the basic attributes of plants and animals. He showed that if a freshwater polyp, or *hydra,* is cut into pieces, each of the parts will grow into a complete new hydra. He also showed that certain simple animals can reproduce by asexual budding and was the first to observe cell division.

Van Beneden, Pierre-Joseph (1809–1894), Belgian zoologist and paleontologist. After studying under Cuvier he specialized on parasites and cetaceans. He is best known for his work on tape worms.

Vicq d'Azyr, Félix (1746–1794), a French physician, is said to be the founder of comparative anatomy and the first proponent of the theory of homology.

Index

W

X

Z

FSC
www.fsc.org
MIX
Papier | Fördert
gute Waldnutzung
FSC® C083411

Kolibri 360 GmbH
Ferdinand-Jühlke-Straße 7
99095 Erfurt, Deutschland
produktsicherheit@kolibri360.de